Tracing the History of Eukaryotic Cells

THE ENIGMATIC SMILE

BETSEY DEXTER DYER

ROBERT ALAN OBAR

COLUMBIA UNIVERSITY PRESS
NEW YORK

Columbia University Press
New York Chichester, West Sussex

Library of Congress Cataloging-in-Publication Data

Dyer, Betsy Dexter.
Tracing the history of eukaryotic cells : the enigmatic smile /
Betsy Dexter Dyer, Robert Alan Obar.
p. cm. — (Critical moments in paleobiology and earth
history series)
Includes bibliographical references (p.) and index.
ISBN 0-231-07592-8 (cloth : acid-free paper). —
ISBN 0-231-07593-6 (paper)
1. Eukaryotic cells—Evolution. 2. Symbiosis. 3. Evolution
(Biology) I. Obar, Robert. II. Title. III. Series.
QH371.D93 1993
574.87—dc20 93-26798
CIP

Casebound editions of Columbia University Press books are printed on
permanent and durable acid-free paper.

Printed in the United States of America
c 10 9 8 7 6 5 4 3 2 1

To Alice Linnea Obar who, during her first three years of life, was exceptionally cooperative and illuminating while her parents wrote this book.

CONTENTS

FOREWORD

Jan Sapp

It was not along ago that the themes discussed in this book were regarded as remote from the mainstream of biological inquiry. Indeed, they were often dismissed as metascientific speculation of little value. Yet as Betsey Dexter Dyer and Robert Alan Obar convey with such clarity and confidence, over the past twenty-five years the evolution of eukaryotic cells and the role of symbiosis in that process have emerged as a vibrant scientific research program. In effect, this book provides ample testimony to the assertion of Theodosius Dobzhansky that "nothing in biology makes sense except in the light of evolution." At the same time, it demonstrates how evolutionary speculation has served to raise new questions for empirical investigation and how the study of cell evolution has become integrated in, and has itself served to integrate, once disparate specialties.

The authors show how information won from the fossil record, genetics, and the most recent techniques of molecular evolution are brought to bear on various facets of cell evolution. They offer an excursion into the study of the evolution of metabolism, sex, and phylogenetic classification, as well as the consensual acceptance that symbiosis has played a key role in the greatest discontinuity in all of organic evolution—that between the prokaryote and the eukaryote.

The approach of the book is exploratory rather than dogmatic. It offers a balanced account of contemporary controversies while revealing the creative suggestions that continue to bring life to the study of cell evolution. This is a book that can and should be read by the widest audience; it offers many questions and puzzles for further research and does so with a sensitivity to description and excitement of the pursuit that will capture the interest of anyone with a concern for fundamental problems of biology.

York University
Ontario, Canada

ACKNOWLEDGMENTS

We could not have produced this book without the help of Kathleen Rogers, who typed the entire manuscript onto the Macintosh and then proceeded tirelessly and even cheerfully through numerous revisions.

The reference librarians at the Wheaton College library, Marcia Grimes and Sherry O'Brien, along with Lyman Green and Marcela Aquilar, searched the world for us through interlibrary loan, to obtain obscure reprints and out-of-print books. It was as though we had access to a major research library, and perhaps it was even better because of the personal attention of the Wheaton staff. The on-line computer search system, established by former head librarian Sherry Bergman, was a luxurious and then indispensable tool for our research.

Lydia Carswell graciously provided child care during some of the crucial summer months when we were writing. And we are much indebted to our regular child care provider, Florence Poholek.

ACKNOWLEDGMENTS

Work-study students at Wheaton College—Jennifer Palaia, Stephanie Jensen, Ongkar Khalsa, and Keven Howe—helped by making copies, tracking down references, and indexing.

Critical readings of the manuscript by Dennis Searcy, Andrew Knoll, and David Bottjer were very helpful in our revision process. Many researchers listed in the final chapter responded generously to our request for ideas about future research directions, while David Bermudes and John Stolz gave ongoing advice and comment. Science writer and copyeditor Connie Barlow made many helpful and thoughtful suggestions that greatly improved our manuscript.

Carolyn Nichols provided many of the pen and ink drawings.

Finally, our former Ph.D. adviser, Lynn Margulis, is a major inspiration for our work. For her many contributions to our writing and research, we thank her.

TRACING THE HISTORY OF EUKARYOTIC CELLS

I

Fossils and Molecules: An Introduction

THIS IS a book about the evolution of complex, eukaryotic cells—that is, nucleated cells rich in subcellular organelles that carry out specific functions, like respiration, photosynthesis, and internal and external cell motility. All plants, animals, fungi, and protoctists (like amoebae and slime molds) are composed of eukaryotic cells. They, rather we, are all eukaryotes. The emergence of eukaryotic cells from prokaryotic, bacterial forebears happened midway between the origin of life and the appearance and explosive proliferation of multitissued animals. It is thus one of the major mileposts in the history of life—indeed, in the history of earth.

Cells are not a part of our macroscopic world and are often difficult to study and interpret even with the best equipment. Tracing the history of cellular evolution through the fossil record is even more problematic. Fortunately, the structure and dynamics of living cells today preserve some of the record of the past. The evidence is tantalizing—and enigmatic. Evolutionary biologist David C. Smith

(1979:128) made this point most eloquently, and we owe the title of our book to his metaphorical insight.

> In nonliving habitats, an organism either exists or it does not. In the cell habitat, an invading organism can progressively lose pieces of itself, slowly blending into the general background, its former existence betrayed only by some relic. Indeed one is reminded of Alice in Wonderland's encounter with the Cheshire Cat. As she watched, it "it vanished quite slowly, beginning with the tail, and ending with the grin, which remained some time after the rest of it had gone." There are a number of objects in a cell like the grin of the Cheshire Cat. For those who try to trace their origin, the grin is challenging and truly enigmatic.

In tracking the emergence of the eukaryotic cell, one enters a kind of wonderland where scientific pursuit leads almost to fantasy. Cell and molecular biologists must construct cellular worlds in their own imaginations. For some researchers, the microscopic and the inaccessible have become so familiar that they are almost palpable. Conversations among cell biologists can be as richly and precisely descriptive as if that knowledge were gained directly, through the five senses. One can temporarily forget that the topic is (for example) nuclear pores, structures so tiny that they have never been visualized directly let alone sensed on a macroscopic level. This kind of intimacy with the subcellular world has perhaps its most famous expression in Evelyn Fox Keller's biography of Barbara McClintock, *A Feeling for the Organism* (1983:117). There, the Nobel laureate reflects on her relationship with her research subject, maize, and how it led to startling discoveries in genetics.

> I found that the more I worked with [chromosomes] the bigger and bigger [they] got, and when I was really working with them I wasn't outside, I was down there. I was part of the system. I was right down there with them, and everything got big—I even was able to see the internal parts of chromosomes—actually everything was there.

This sort of facility, the ability to move about with ease in a microscopic world, must be cultivated. We hope that our treatment here of the cellular world, and particularly its breathtaking history, will nurture the knowledge and the enthusiasm of those who also wish to make the imaginative leap. For those who have already done so, we hope this book will reinforce and renew the vision.

Imagination, to some degree, is essential for grasping the key events in cellular history. So many of the components found in cells retain cryptic remnants of the past. It is as if cells are enthusiastic collectors of souvenirs. Their attics and basements are full of tokens of the past, but most of the items are broken or so outdated that their original functions are no longer obvious. Moreover, some of the souvenirs can replicate and mutate autonomously, filling the attics and basements with a multitude of surprises. Lynn Margulis and Dorion Sagan have made this point in extending the metaphor of David C. Smith. "It is as if the Cheshire Cat has had a litter of kittens that are playing everywhere, yet more enigmatic and faded than their ancestor" (1986:105).

Many of these replicating entities are remnants of what were once distinct symbionts (intimately associated organisms). Symbiosis is, in fact, the primary force underlying the evolution of complex cells. It is thus a major theme of this book.

Symbiosis is not a rare occurrence. It is an exceptional organism that lives its life alone, unassociated with other species. Associations and interactions between organisms are the rule. In fact, the rare solitary individuals are apt to exist as laboratory artifacts—"axenic" organisms maintained at considerable expense with a sterile supply of food, water, and air. And not only are axenic organisms difficult to keep, but they do not thrive compared with their networked wild counterparts.

Some of the most intimate associations between species are called symbioses. Such relationships may include situations in which one organism is attached to another, wrapped around another, or even fully inside the other. Interactions may range from the casual sharing of food to very specific exchanges of nutrients between partners. Some symbiotic companions are even genetically intimate, exchanging and sharing genes that code for structures used in common.

Whether in casual contact or intimate symbiotic bond, living associates are a fundamental part of the process of natural selection. The fitness of an organism may be profoundly affected when selection acts on a partner species. Organisms in association thus shape each other and evolve together.

In our view, the most striking example of the role of symbiosis in the history of life is the origin and evolution of complex, eukaryotic cells. The theory that complex cells might actually represent long-standing and highly evolved symbiotic associations among simple cells was described in detail in 1967 by Lynn Margulis (then Lynn Sagan) in the *Journal of Theoretical Biology.* Margulis depicted complex cells not as individuals but as communities of simpler (bacterial) cells. The theory was at first strongly disputed. But as corroborating evidence accumulated (Margulis 1981), criticism diminished. Now, twenty-five years later, the theory is well established; it is standard fare for introductory level textbooks.

As is not uncommon in intellectual history, the idea that complex cells arose through symbiosis, "symbiogenesis," has multiple origins in time and place. An American, Ivan E. Wallin (1923), and a Russian, Konstantin S. Mereschkovsky (1910), both built their ideas of the symbiotic origin of eukaryotic cells on considerable work (see Taylor 1987). But Wallin and Mereschkowsky were too far ahead of their times and the techniques for testing their hypotheses were then crude, so both were forgotten for decades.

The story of the origin of complex (eukaryotic) cells begins at about two and a half billion years ago (bya) when the Earth was teeming with bacterial life in the waters and moist soils, and even on rocks and desert soils. Life may indeed have been teeming, but by human standards the plenitude would not have been obvious. A human observer would have noticed only slight pigmentations in the soil and water and perhaps some of the distinctive smells of active bacterial metabolism (figure 1.1). At this time (2.5 bya) bacteria had been around for about 1.5 billion years and had, with surprising rapidity, evolved most of the forms of metabolism present today.

The question of which metabolism came first has not, however, been settled. We will support (and develop in later chapters) a view that fermentation was first. Fermentation is a type of heterotrophy in which organisms exploit for energy and nutrients simple sugars and other molecules that formed spontaneously on the anoxic, early earth. Fermentation is still a viable lifestyle (and a valuable enterprise for makers of bread and beer), but today the sugars and other molecules are generated not spontaneously, but by other organisms. Later, other bacterial metabolisms began to evolve. Through photosynthesis (autotrophy) sugars and other food molecules could be generated from the planet's vast reserves of carbon dioxide, plus some source of hydrogen, using sunlight as a source of energy. Because they could create their own food from simple building blocks, autotrophic organisms had a distinct advantage and got an early and firm foothold among the evolving organisms.

Early in the evolution of autotrophic metabolisms some bacteria found a way to tap the wealth of hydrogen making up part of water molecules. Stripped of hydrogen, the oxygen in those water molecules was a powerful reactant, oxidizing iron and other metals in the soils, hydrocarbons, and even living organisms that had not evolved membranes strong enough to withstand the ravages of free radicals emitted by their neighbors. Eventually, oxygen began to accumulate in the atmosphere.

Until this point the atmosphere had been nearly or entirely free of oxygen, which was fortunate because during the origin of life the evolving molecules composed of hydrogen, carbon, and nitrogen would have been at great risk of decomposing by oxidation. Later, as oxygen slowly accumulated in the atmosphere, organisms gradually

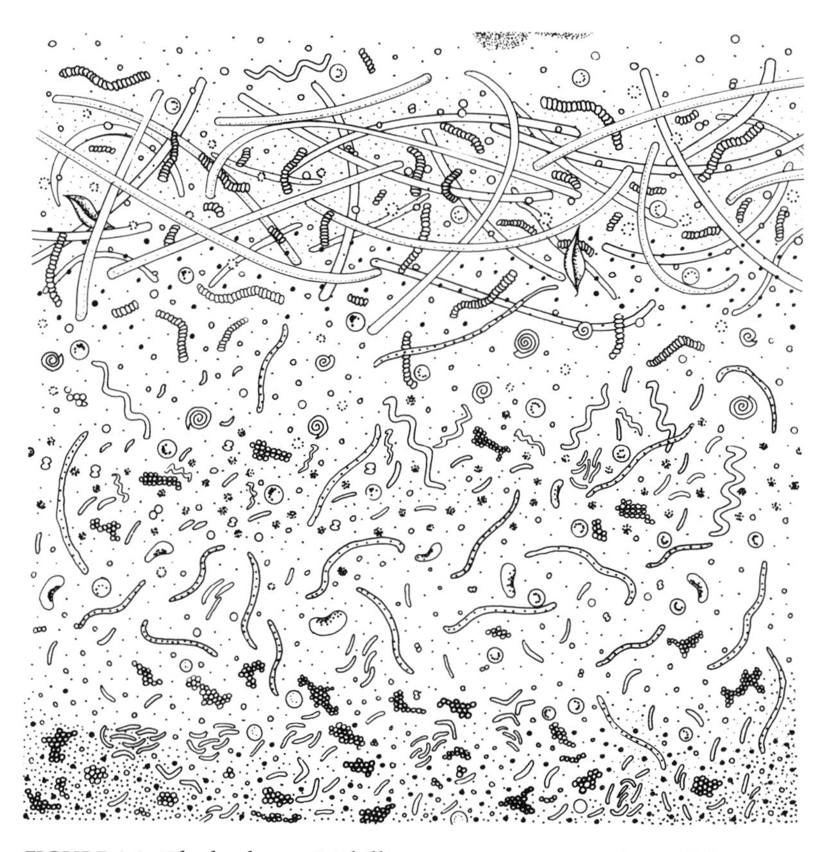

FIGURE 1.1. The landscape 2.5 billion years ago was teeming with bacteria. Drawing by C. Nichols.

evolved cellular mechanisms to cope with its toxic effects. One of the more innovative mechanisms was a system that turned a burden into a blessing. It was a new and highly efficient heterotrophic metabolism: respiration. Respiration was not only an effective method for removal of toxic oxygen, but it put oxygen to good use at the end of a pathway for metabolizing food molecules. In fact, so effective was this new metabolism that organisms (such as ourselves) that utilize it tend to think of free oxygen as an entirely beneficial gas; but even respiring organisms require oxygen detoxification mechanisms to protect their essential molecules.

And so by 2.5 bya, a veneer of life had turned earth's surface into a pigmented landscape consisting of colonies of bacteria metabolizing in all sorts of ways, including various modes of fermentation, respiration, and autotrophy. From then on, the atmosphere would gradually accumulate its full complement of oxygen, a major pollution event on the part of those early photosynthesizers but one which had been dealt with effectively by many bacteria. Those bacteria that remained susceptible to the toxic effects of oxygen either went extinct or restricted their ranges to places bereft of oxygen either because of the structure of the sediments or the oxygen-consuming habits of a buffer zone of neighbors.

The diverse communities of bacteria at 2.5 bya had not only evolved the major metabolisms found on earth today, but they had also invented most kinds of gene exchange, such as mechanisms for the "horizontal transfer" of DNA across species boundaries. The only major form of genetic exchange yet to be evolved was the complicated kind of sexual exchange invented by eukaryotes. Thus, at 2.5 bya, the stage was set for an extraordinary sequence of events—the acquisition by a host bacterium of at least two, and possibly three or more symbionts, and the evolution of complex cells.

MICROFOSSILS: THE CHALLENGE OF INTERPRETATION

THE FOSSIL record is an obvious place to search for clues about the evolution of eukaryotic cells. However, as with much of the fossil record, interpretation is often difficult. The evidence itself can be patchy, poorly preserved, and ambiguous. Evidence may not exist at all for most tiny, delicate cells and their microscale metabolisms. In this chapter, we will review the fossil evidence concerning the early evolution of eukaryotic cells and will conclude, on the basis of direct and indirect evidence, that the eukaryotic cell began its evolution about 2.5 bya.

When cells die they are usually degraded rapidly by organisms in the environment, leaving no trace. Degradation may be slower in environments with little oxygen, high acidity, or high salinity. But even then individual cells do not preserve particularly well as microfossils in most sediments. This is especially true for carbonate sediments, the large crystals of which crush delicate cells. Microfossils may be preserved if the carbonates are replaced early in diagenesis with chemically precipitated silica, which forms microcrystals (Schopf and Walter 1983; Knoll 1985). The process of mineralization is something like that which forms petrified wood, in which silica permeates and preserves the plant tissue.

Sedimentary rocks mineralized with microcrystalline silica are called cherts. Black cherts, which are black owing to remnants of organic compounds or iron sulfide, are even more likely to have microfossils. Cherts formed from sediments that were originally in saline, carbonate-rich areas also preserve microfossils well (Walter 1983; Hoffman and Schopf 1983). However, even well-preserved microfossils may be deceptive. The sheaths and envelopes of bacteria may preserve well, and may actually be all that can be seen in a particular microfossil. What appear to be internal structures may actually be collapsed walls and membranes of the cell that was once inside. For example, an attempt to do transmission electron microscopy of microfossils purported to be eukaryotic algae from the Bitter Springs formation (1 bya) in Australia, yielded some rather empty looking photographs of cells, difficult to interpret (Oehler 1976).

Size, too, can be deceptive. A large, round cell may have enlarged under osmotic pressure. A smaller, folded one may have been compacted by sediments (Hoffman and Schopf 1983). And so with this cautionary introduction, here is a review of some of the more important microfossil discoveries and how they might be interpreted.

The earliest rocks bearing microfossils are about 3.5 billion years old and are from the Onverwacht formation in South Africa and the Warrawoona formation in Australia (Schopf 1992). These fossils are small spheres and filaments and have been interpreted to be prokaryotes (bacteria). The shapes and sizes give few clues as to their exact identity and what their metabolism might have been like. However, the possibility that they were photosynthesizers is supported by other types of evidence found in other rocks of the same age (chapter 2). Prokaryotic microfossils appear with consistency in appropriate sedimentary rocks from 3.5 billion years to the present (represented by prelithified sediments). The question of interest is at what point in the geologic record is there reliable evidence for the first eukaryotic cells?

Unfortunately, there is no simple answer to this question. The reason is that cells with tough outer coverings (sheaths, walls, and envelopes) or with mineralized skeletons (such as the silica tests of diatoms) fossilize best. Furthermore, environments in which dead cells are degraded very slowly are more likely to produce microfossils. None of these conditions may pertain to the earliest eukaryotes. The first eukaryotes had cell membranes, but they probably lacked the enveloping protection of sheaths or walls manufactured by prokaryotes. In addition, they probably evolved in hot, acidic water.

Fossil evidence of internal structures may be deceptive as indicators of a eukaryotic cell as they may represent nothing more than the crumpled remains of a bacterial cell that has pulled away from the tougher envelope or sheath (Hoffman and Schopf 1983). A browse through a comprehensive handbook like *The Prokaryotes* (Starr et al. 1981) quickly indicates that complex looking cells in multicellular arrangements may be found in many prokaryotes. For example, the cyanobacteria *Chroococcus* and *Aphanocapsa* are found in clusters of four cells; they can look deceptively like eukaryotic cells that have just completed meiosis (see figure 1.2). Myxobacteria can make large, complex, multicellular "fruiting" structures and many bacteria such as *Bacillus* produce internal spores. It is therefore with caution that researchers propose a particular microfossil to be an early representative of a eukaryotic cell.

Size may be a legitimate criterion (along with morphology and the frequency distribution of size for eukaryotes), as eukaryotic cells in general are larger than prokaryotic ones. The large sphere-shaped microfossils found in some sedimentary rocks from China 1.8–1.9 bya, or the large filaments from rocks 2.1 bya in Michigan may be among the first fossilized eukaryotes. Unsegmented, filamentous fossils are also found at 1.4 bya in Montana and China. At 1–2 mm wide and 130 mm long these are considered to be the first megafossils (Walter et al. 1990).

After these earliest occurrences (2.1–1.8 bya) and in particular

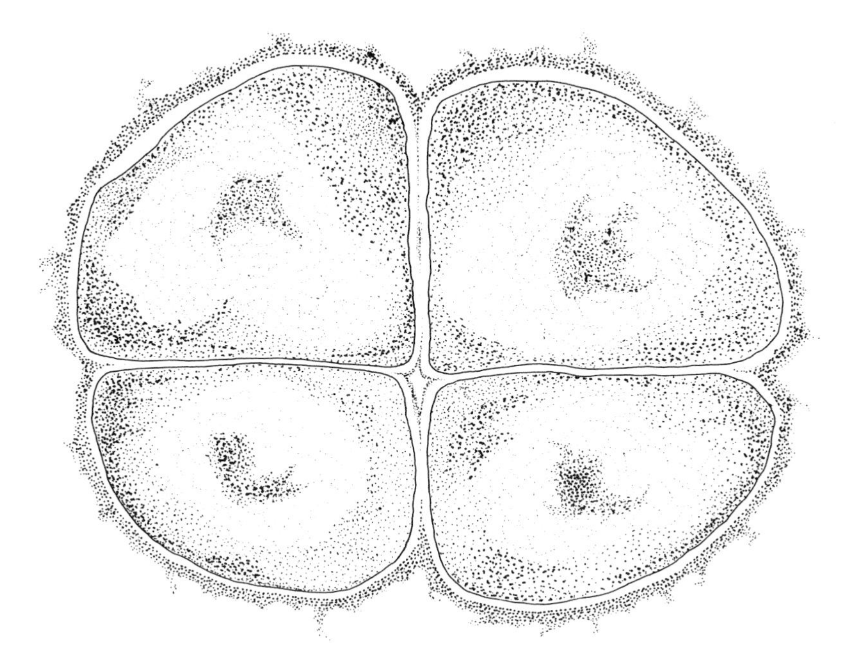

FIGURE 1.2. The fact that prokaryotic cells may sometimes be found in clusters of four (tetrads), as these are, suggests that tetrads are not good indicators of eukaryotes in the fossil record. It had been thought previously that such tetrads might be the typical four products found after a type of cell division characteristic of eukaryotes, meiosis. Drawing by C. Nichols.

after 1.1 bya, there seems to be a general increase in the size of fossil cells to that which is clearly "eukaryotic" (Schopf and Oehler 1976; Mendelson and Schopf 1992). The interpretation may be marred, however, by the general scarcity of fossils before 2 bya. It has been suggested (Horodyski 1980; Vidal 1984) that some large microfossils found in the Chamberlain Shale of Montana (1.4 bya) might be among the earliest fossilized eukaryotes, based mostly on considerations of size. Despite the aforementioned uncertainties and drawbacks, there are two reasons that using size as a criterion may be reasonable. First, most extant prokaryotes do have very small cells (Starr and Schmidt 1981). The exceptions are usually long filamentous bacteria which, in spite of lengths up to 100 micrometers (μ) for *Achromatium oxaliferum,* have diameters of just 5 micrometers. There are only a few, exceptional bacteria that grow as large as the smallest eukaryotes. *Beggiatoa gigantea* has cells with dimensions of 26–55μ × 5–13μ, and *Clonothrix fusca* has cells that are arranged in trichrome filaments. *Cylindrospermum* is 10 × 33μ and *Macromonas morilis* may be 6–14 × 10–30μ. The behemoth bacterium *Epulopiscium fishelsoni* can be 80 × 600μ (more than half a millimeter), in the digestive systems of fish (Clements and Bullivans 1991). None, however, approach the 1–2 mm thickness of fossil filaments from 2.1 and 1.4 bya.

In general, bacteria have sizes in the range of 1 μ by 1 μ, which is thought to be the most efficient package for the movement of metabolites in a cell, and the most efficient surface-to-volume ratio for cells that do not have extensive internal membranes or a cytoskeleton. An increase in size is thus exactly what would be predicted for a group of prokaryotes making the transition to eukaryotes by acquiring symbionts. Evidence from studies of rRNA (ribosmal RNA) indicates that the first eukaryotes probably arose from prokaryotes that had acquired certain proteins that could then be used for a cytoskeleton and a system of internal membranes (to be discussed later). One of the consequences that could be most easily seen in the fossil record would be an increase in cell size as cell architecture becomes more stable and more efficient.

Changes in cell architecture as evidenced by an increase in cell size seem to have been closely followed by the acquisition of mitochondrial and plastid symbionts. The acquisition of symbionts often results in hypertrophy (increase in size) of the host, and thus changes in size may be a good indicator of symbiotic events as well as changes in cell architecture. Therefore, Vidal and others may be correct in citing a parameter as simple as size for the first evidence of eukaryotes in the fossil record. In any case, a paucity of other types of evidence, forces interpreters of the fossil record to concentrate on size.

THE FIRST EUKARYOTIC MICROFOSSILS

THE EXTINCT *Grypania* is the earliest known fossil that may be identified as a eukaryote. It is enormous: up to 1.5 mm thick and 30 mm long (in its large form) and about 1 mm thick and 90 mm long (in its thinner form). It has been found in 2.1 bya rocks in Michigan. Unbranched and unsegmented, *Grypania* grew in coils, held by thick and rigid cell walls (Han and Runnegar 1992).

Large (40–200 μ) spherical microfossils 1.8 to 1.9 billion years old found in sedimentary rocks from China are also among the oldest examples of microfossils identified as eukaryotes (Zhang 1986; Knoll 1992). Rocks of 1.69 bya from several sites, including in China, supply additional evidence that eukaryotes lived at this time. These rocks contain steranes chemically derived from molecules such as steroids, which are found almost exclusively in the membranes of eukaryotes (Hoffmann and Chen 1981; Summons and Walter 1990; Knoll 1992). Thus, it can be rather conservatively concluded that between 2.1 and 1.7 billion years ago, eukaryotes of unknown taxa were a part of at least some microbial communities. However, it is highly unlikely that these sterane-associated fossils represent the very first eukaryotes.[1]

The availability and distribution of sedimentary rocks containing microfossils is extremely patchy before 1.0 billion years ago (Knoll 1992). While the presence of microfossils might be taken as evidence for the existence of certain organisms, the absence of microfossils may signify nothing but bad luck in sampling and difficulties with interpretation. This is especially true for sedimentary rocks dated at or a little older than 1.9 billion years, such as the Gunflint cherts of Canada and their famous assemblage of microfossils. Representing a remarkably diverse and distinctive community, the Gunflint cherts were discovered by Barghoorn and Tyler in 1965. Since then, they have been intensively analyzed and interpreted, especially in regard to the question "Are there any eukaryotes?" The problem here is not so much a lack of microfossils of sufficient size and complexity but the correct interpretation of those microfossils. For example, *Eosphaera* (24–115μ in diameter) has the appearance of eukaryotic green alga in colonial form, something like a modern day *Volvox*, according to Kazmierczak (1979). *Eosphaera*, along with *Huronospora* (7–24μ diameter), has also been interpreted as red algae of the ancient bangiacean group (Tappan 1976). Other researchers, however, prefer to call these and other Gunflint taxa problematic in that they do not have clear identities as either prokaryotic or eukaryotic (Licari and Cloud 1968; Knoll 1992). For reasons set forth at the beginning of this chapter, neither the size nor the apparent complexity of many of the Gunflint microfossils precludes them from prokaryotic status.

[1]The first eukaryotic cell identified to a modern taxon is believed to be a bangiacean red alga of 1.26–0.95 bya; see Butterfield et al. 1990.

ESTIMATING A DATE FOR THE FIRST EUKARYOTES

THE EVIDENCE is thus suggestive that somewhere between 2.1 and 1.7 bya identifiable eukaryotes had begun to enter the fossil record. But were these first identifiable eukaryotic fossils actually the first eukaryotes?

The tough, well-preserved walls and sheaths of the first eukaryotic fossils are highly suggestive of photosynthesizers such as (plastid-bearing) algae. As there are no photosynthetic eukaryotes lacking nuclear membranes or mitochondria, it is believed that plastids were acquired as photosynthetic symbionts only after the evolution of the cytoplasmic components and after the acquisition of mitochondria (as respiratory symbionts). Thus the first widely recognized fossils of complex cells are probably fairly advanced eukaryotes, with photosynthetic capacities. The 2.1–1.7 bya dates for these fossils thus bracket a rather late stage of eukaryotic evolution. Can time intervals for the earlier events in eukaryotic development also be established in spite of the dearth of microfossils?

One constraint may be a significant event in the evolution of the environment, the accumulation of oxygen in the atmosphere. Early in the history of prokaryotes (perhaps as early as 3.5 billion years ago) oxygen-generating photosynthesis evolved. The highly reactive oxygen readily bound with minerals and hydrocarbons, but free oxygen also began to accumulate locally in patches of soils and as concentration gradients near photosynthesizers in aquatic and marine habitats, and finally in the atmosphere and oceans at large. By 2.8 to 2.4 billion years ago or earlier trace amounts (probably less than 0.5 percent) of oxygen had accumulated in the atmosphere—possibly sufficient to support aerobic respiration (Knoll 1992; Klein 1992; Kasting et al. 1992). This interval of 2.8 to 2.4 billion years thus provides an early bracket for the existence of aerobic bacteria and possibly for the immediate ancestors of mitochondria. Furthermore the putative host for mitochondria was probably also aerobic or could at least deal with the toxic effects of oxygen, as evidenced by the extant capabilities of the eukaryotic cytoplasm. Although it may be impossible, with current evidence, to fix an exact date for the acquisition of respiring organelles by eukaryotic ancestors, a reasonable time bracket might well be 2.8 to 2.4 billion years ago. Thus, 0.3 to 1.1 billion years might have elapsed between the acquisition of respiring organelles and the later acquisition of photosynthesizing organelles (see figure 1.3).

Knoll (1992) and others suggest that most of the groups of photosynthetic eukaryotes evolved no earlier than about 1 bya. At that point there was an extensive radiation. This evolutionary spurt is evident, for example, in the complex community of organisms at Bitter Springs Australia (1 bya). The assemblage seems to include several types of eukaryotic algae (Barghoorn and Schopf 1965; Schopf 1968).

WHY MITOCHONDRIA BEFORE PLASTIDS?

IT MAY be that mitochondria are necessary for certain activities of plastids. One example (explained further in chapter 7) is the role of mitochondria in processing a plastid waste product, glycolate, and returning it in usable form to the plastid. Oxygen toxicity may be another reason for the acquisition of plastids only after mitochondria were established. Oxygen, produced as a waste product by plastids, could be most efficiently detoxified in mitochondria which use oxygen at the end of the electron transport pathway (see chapter 6).

Yet another reason for concluding that mitochondria arose first has been proposed by Knoll (1992). The concentrations of oxygen in the atmosphere earlier than about 1.9 billion years ago would not

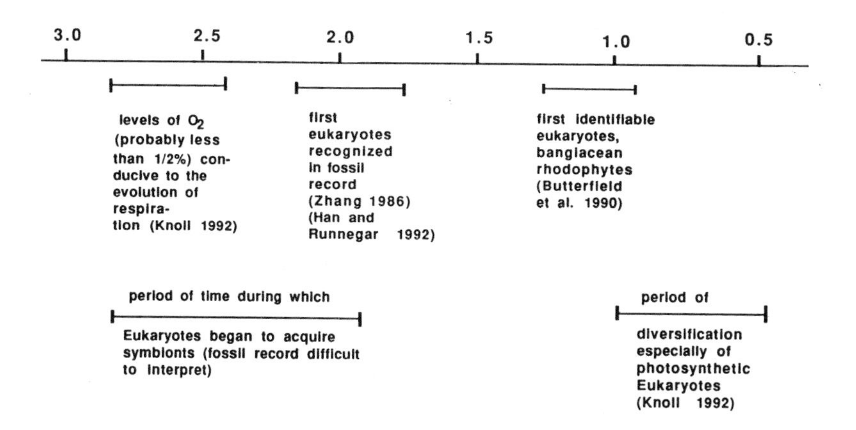

FIGURE 1.3. Events in the geological history of early eukaryotes (in billions of years before present).

have been conducive to nitrate production (an important source of nitrogen for photosynthesizers). Although some photosynthetic prokaryotes compensated by fixing their own nitrogen (from atmospheric N_2), others probably scavenged for scarce nitrates. Knoll suggests that when oxygen levels and consequently nitrate levels increased, conditions were more conducive to the establishment of symbioses. Then too, the rise in nitrogen posed problems for nitrogen-fixing prokaryotes, as highly reactive oxygen disrupts the delicate process of splitting nitrogen molecules. One possibility for refuge is within the cells and tissues of eukaryotes (as in the nitrogen-fixing root nodules that legumes provide for their symbionts today). Whether or not this is the explanation, it certainly is striking that during the period in which oxygen began to increase significantly in the atmosphere, recognizable eukaryotic fossils (apparently of photosynthesizers) appear in the fossil record.

ESTIMATING A DATE FOR THE ORIGINS OF THE HOST CELL

A SOMEWHAT more challenging problem is that of ascertaining the date of origin for the host cell, the pre-eukaryotic prokaryote that acquired respiring and photosynthesizing symbionts, respectively, between 2.8–2.4 bya and 2.1 bya. The problem impinges not only on paleontology and molecular biology but also semantics. The pre- or early eukaryotic cell (described further in chapters 4 and 5) is hypothesized to be either completely or entirely lacking a cell wall (and thus a fossil marker) and it probably had some kind of network of internal membranes (also not fossilized). The most significant of these internal membranes is the nuclear membrane (envelope); the nuclear membrane is, in fact, a defining characteristic of eukaryotes. Indeed there are groups of eukaryotes completely lacking in either mitochondria or plastids but which have the internal membranes (and certain other capabilities) that establish them as eukaryotic. These groups, such as diplomonida, will be further discussed in chapter 4, particularly with attention to the question of whether mitochondrial absence is a primary characteristic or represents a secondary loss.

Whether or not extant groups like diplomonida represent a very early eukaryotic lineage, it is reasonable that such groups must have existed at one time, perhaps just before the acquisition of mitochondria. If the advent of a well-defined enclosure for the genetic substance, a nuclear membrane, marks the initial transition from prokaryotic to eukaryotic, then it may be possible to approximate a date for this event. Surprisingly, the date may fall within the range already established in the previous sections for the acquisition of mitochondria and plastids (between 2.1 and 2.4–2.8 billion years ago). This overlap suggests that the increase in oxygen levels during this period (which was indirectly conducive to the acquisition of symbionts) may also have selected for nuclear membranes.

DNA is especially vulnerable to the toxic properties of oxygen (particularly in its free radical or reactive forms). Control of oxygen toxicity may well have been a major selection pressure for the evolution of a nuclear membrane (e.g., Margulis and Sagan 1986). The nucleus forms an anoxic and relatively safe environment for the DNA, but not a perfectly safe one. For example, it has been noted that DNA closer to the nuclear membrane and to the relatively oxygen-rich cytoplasm is more susceptible to damage (Kvam et al. 1990). Also, mitochondrial DNA—that is, DNA found within the oxygen-respiring organelles of eukaryotes—is especially at risk to oxygen toxicity. Over the course of a mitochondrial life span this may have profound degenerative effects (Gupta et al. 1990). Indeed, there are hypotheses that the degeneration of aging cells may owe, in part, to oxygen damage to mitochondrial DNA.

Thus it is not unreasonable to argue that the period of time in which oxygen accumulated and in which mitochondria and plastids were acquired was also the time in which the nuclear membrane evolved. Margulis and Sagan (1986) hypothesized that the very mechanisms necessary to evolve a flexible membrane system and nucleus might also have been dependent upon a significant rise in oxygen. The biochemical pathway by which steroids are made requires oxygen; steroid derivatives are found almost exclusively in eukaryotic membranes and some have the characteristic of making these membranes more flexible. Thus steroid synthesis may have evolved during the time when oxygen was first available for such synthesis. Steroid synthesis might initially have served as a way to render harmless those oxygen molecules that found a way in to the cell or that were generated by the plastids; the role of steroids in rendering membranes more flexible might have been a secondary discovery—the kind of fortuitous circumstance called "preadaptation."

There is thus substantial (if indirect) evidence for arguing that the nucleus evolved at nearly the same time or shortly before the acquisition of mitochondria and plastids. But this does not entirely solve the genealogical problem. The bacterial cell line that was destined to become the pre-eukaryotic cell may have diverged from the other bacterial lineages considerably earlier than 2.8 to 2.4 billion years ago. Fossil evidence to support such a claim is lacking. However, molecular evidence (which will be discussed later) does suggest a very early branching date.

Whether or not this plausible early branching pushes back the date of origin for the first eukaryotes is actually a semantic problem and one that has not been resolved. Do these early pre-eukaryotic cells, which probably lacked a membrane-enclosed nucleus and which almost certainly lacked mitochondria or plastids, qualify as eukaryotes on the basis of events that would occur in their future? For purposes of this book, our answer is no. We have chosen to use a definition of eukaryotes that includes the presence of a eukaryotic-type of nucleus. The distinction is an important one (especially since the nucleus is a primary identifier of all eukaryotes today), and so we consider those earlier cells to be pre-eukaryotic prokaryotes. Their origin, based upon molecular data may be as early as 3.5 bya. Their divergence from the other branches of bacteria is the topic of a later section in this chapter.

AN ESTIMATED DATE FOR THE ORIGIN OF MOTILITY ORGANELLES

MORE CONTROVERSIAL than the origin of plastids and mitochondria is the origin of organelles responsible for some types of motility, including tubulin-based structures. (Details of the controversy are described in chapter 8.) Whether or not these organelles are of symbiotic origin, it may be possible, using some of the indirect evidence developed in this chapter, to extrapolate a date of origin for them.

First, the problem of fossilization of tubulin-based motility organelles and related structures such as the spindles used in eukaryotic cell division is similar to that encountered with other delicate cell structures. That is, there may be no direct fossil record at all. All eukaryotes known today have some sort of tubulin-based motility system or what appears to be remnants of such a system. Tubulin is a protein making up the tubelike structure that forms the basis for a diversity of structures, including motility organelles (e.g., cilia).

Tubulin-based structures are as universal an identifier of eukaryotes as the nucleus is. Tubulin structures, such as mitotic spindles, seem to be almost obligatory for eukaryotic cell division. They also form part of the cytoskeleton typical of eukaryotic cells. A cytoskeleton may be requisite architecture for cells that tend to be orders of magnitude larger than prokaryotic cells. The near universality of some of the tubulin-structures suggests that they may have been established or already were established during the period hypothesized for the origin of eukaryotes (2.8 to 2.1 bya). The large size of eukaryotic fossils found at 1.7–1.9 bya may have been facilitated by microtubules in the cytoskeleton. The presence of a nucleus, which may have originated as early as 2.8–2.4 bya, seems to demand of most eukaryotes a tubulin-based system (spindles) for DNA division. The three tubulin proteins, alpha, beta, and gamma, are some of the most conserved and universal proteins in existence.

It is reasonable, then, to conclude that by the time that large cells with defined nuclei evolved, the equipment for intracellular structure and motility had already been acquired by the earliest eukaryotes. But how much earlier might the intracellular structure and motility have arisen? Unlike the previously discussed stages of eukaryotic development, there is no reason to link tubulin evolution with a rise in oxygen levels. Anaerobic eukaryotes lacking both mitochondria and plastids are found today in the anoxic environment of (for example) termite intestines. But these eukaryotes do have motility organelles. The safest conclusion may be not to estimate the earlier appearance of motility organelles but to say merely that these organelles were established by the time the rest of the eukaryotic cell was attaining its identity, that is, evolving the nucleus and acquiring mitochondria and plastids.

CONCLUSIONS REGARDING A DATE FOR THE ORIGIN OF EUKARYOTES

BASED UPON the previous arguments extended from fossil evidence and summarized in figure 1.3, a rather intriguing conclusion is that eukaryotes evolved in a hurry (geologically speaking). That is, during a critical 0.7 to 0.3 billion year period (sometime between 2.8 and

2.1 bya) during which oxygen accumulated precipitously in the atmosphere, the pre-eukaryotic cell was transformed to eukaryotic via the evolution of a distinct nucleus with nuclear membrane and the acquisition of mitochondria and plastids—in that order (but perhaps nearly simultaneously). Motility organelles and related tubulin-based structures may predate these events but it is not clear by how much. An important question and one that will be addressed next is: To what extent are these geologically derived dates confirmed by the data being generated by molecular evolutionists?

DISCERNING THE MAIN BRANCHES OF ORGANISMS FROM MOLECULAR EVIDENCE

THE GENETIC substance, DNA, of all organisms is susceptible to random changes in sequence (mutation) not only because of environmental agents (mutagens), such as radiation and trace chemicals, but also owing to simple, everyday mistakes in replication and repair. The accumulation and inheritance of mutations is a fundamental mechanism in the evolution of organisms. Scientific understanding of that process has also brought about the development of one of the most effective tools for establishing relative, and sometimes absolute, dates for divergences in the tree of life.

One of the basic ways by which the evolutionary phylogeny or family tree for a group of organisms might be reconstructed is to look not so much at the organisms themselves but at the information derived from DNA, RNA, and protein sequences. If we can assume that all the genomes of all organisms experience approximately the same degree of contact with mutagens and the same rate of making mistakes (these assumptions are not without controversy), then by looking at the number and positions of mutations in DNA, it should be possible to determine at what point various organismal lines diverged or formed branches in the family tree. Deriving family trees from sequences is one of the major goals—and challenges—in the field of molecular evolution.

One challenge in using molecular evolution to ascertain genealogies is the issue of alignment. If two linear sequences contain the same or approximately the same number of sequence elements, aligning them is simple, and it can be done without computer assistance. But when pairs (or families) of sequences contain extended deletions or insertions, the act of alignment requires sophisticated computer algorithms that tally "scoring penalties" for the introduction of gaps and the alignment of mismatches. Such tools make possible rapid comparison of large numbers of sequences and the detection of patterns that would otherwise be tedious or impossible to find, but at a price: the results are affected strongly by the assumptions introduced into the comparison, and these assumptions, based on qualitative ideas, must be made quantitative for the computer to make use of them.

Another issue is whether to focus on similarities (sequence matches) or dissimilarities, or to include factors that pertain to both. Sequence dissimilarities are more important in assessing mutation rates and mechanisms, but similarities (unless they owe to convergence) may connote common origin, which is what evolutionary comparisons are all about. The fascination (and the controversy) thus derive from the fact that sequence comparison is both a qualitative and a quantitative field with no clear division between the two methodological modes. It is the wealth of data in this field that particularly lends itself to different interpretations. In analyzing questions for which molecular evolution supplies the sole evidence, the answers are always approximations—and highly debatable approximations, at that.

Which sequences to choose in making such analyses is another problem for molecular evolutionists. Favorable sequences are those that are ubiquitous in the groups of organisms being studied. Also, for ease of analysis it helps considerably if the sequences are relatively short, abundant in the cell, and easy to isolate and purify.

Cytochrome, a small, abundant protein found in nearly all organisms, is an attractive target for protein sequencing. It is considered to be highly conserved because of its functional importance; that is, it has tolerated the slow cumulative effects of mutations over millions of years yet cytochrome molecules in all species are still recognizable as part of the cytochrome molecular family. Cells tend to be intolerant of changes in molecules of great functional importance. Thus by choosing the cytochromes, a molecular evolutionist is assured of seeing a slow and steady change of sequence but enough similarity to permit mapping on a family tree. Throughout the subsequent chapters in this book many conclusions about the relatedness of organisms and their parts have been derived from the literature of cytochrome sequence analyses.

We will concentrate mostly on a different set of sequences, distinguished by their extreme conservation and virtual ubiquity in organisms–the short and plentiful sequences of RNA found in structures called ribosomes, which are essential for protein synthesis in all organisms. These ribosomal RNA (rRNA) sequences are coded for by a family of genes, and either the RNA or the DNA sequence of the corresponding genes may be isolated and analyzed. Most of the studies in molecular evolution aimed at discerning divergences that took place a billion or more years ago use sequences of rRNA. In a previous section, fossil evidence was used to derive an approximate lineage for eukaryotic cells. Here, the molecular evidence for that lineage will be presented and compared, thereby linking two, once disparate fields: paleontology and molecular biology.

THE THREE (OR FOUR) MAIN BRANCHES OF ORGANISMS

ONE INTERPRETATION of early life based on sequences of rRNAs and other conserved molecules—that of Carl Woese and Sidney Fox (1977) and Woese et al. (1990)—claims that the oldest branches in the family tree of all known organisms are threefold (figure 1.4). The earliest divergence splits the prokaryotes into two groups, the Eubacteria (Bacteria) and the Archaebacteria (Archea). At a somewhat later point a second divergence splits the Archean branch of prokaryotes into Archaebacteria and Eucarya—the posited direct ancestors of the Eukaryotes (Woese and Fox 1977; Woese et al. 1990).

According to Woese and colleagues, the Eubacteria and Archaebacteria are the two groups that comprise the extant Prokaryotes. The Eucarya, in spite of their suggestive name, were probably not in their early history recognizable as eukaryotic cells—that is, with a nucleus and other eukaryotic components of the cytoplasm. For a considerable period of time the Eucarya must have strongly resembled their closest deep-branch relatives, the Archaebacteria. They probably did not begin to attain the striking complexity of true eukaryotes until a series of symbiotic events occurred, long after the initial divergence. The early part of the lineage should be considered pre-eukaryotic. Thus, the Archaebacteria may provide a useful comparison group for understanding the pre-eukaryote.

That Eucarya (with no known extant bacterial members) and the extant Archaebacteria have several eukaryote-like features in common, described in chapter 4, suggests deep roots for the evolution of eukaryotes. Significant among these common features is a complete or near lack of a bacterial cell wall; rather, they share a flexible enclosing membrane. The Archaebacteria probably represent the only good comparison group (albeit distant) for pre-eukaryotes short of eukaryotes themselves, since it appears that all pre-eukaryotes either became eukaryotes or became extinct (or are, as yet, undiscovered). Perhaps as symbionts and internal membranes were acquired by the pre-eukaryotes during their transition, any members of this group that did not acquire symbionts and membranes were quickly outcompeted and their lineages lost. The acquired symbionts of

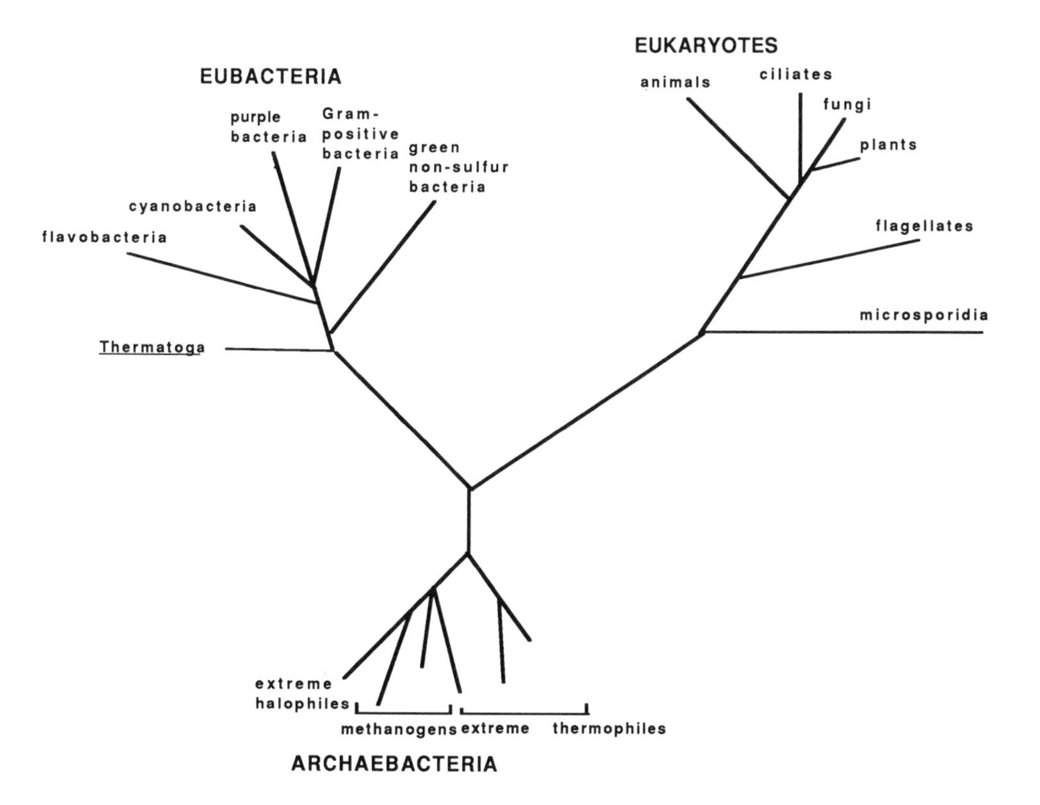

FIGURE 1.4. Woese and Fox (1977) have proposed that there are three major, deep branches on the family tree of all organisms. After Woese and Fox 1977.

Eucarya were, by the way, all Eubacteria—thus reuniting two lineages of ancient divergence.

The tripartite prokaryotic tree of Woese and colleagues is not, however, the only candidate. James Lake (and later in collaboration with Maria Rivera) proposes a four-fold genealogy (Lake 1988, 1989; Rivera and Lake 1992). He focuses on the pattern and occurrence of a conserved sequence of a protein that participates in protein synthesis (elongation factor Tu/1). Lake also used rRNA sequence data but with a different algorithm than that of Woese. He established a family tree that accommodates evidence that eubacteria and eukaryotic sequences seem to evolve ten times faster than archaebacterial sequences. While the actual mutation rate may be the same for both lineages, Lake maintains that the rate at which changes become fixed in the population (the overall evolution rate) is higher in eukaryotes and eubacteria than in archaebacteria. He employed an algorithm in his sequencing work that corrects for this perceived difference.

Lake found four main branches of organisms: the Eubacteria, the Archaebacteria (but modified from that of Woese), the Eocytes (a new group), and the Eukaryotes (see figure 1.5). Specifically, Lake's scheme has split Woese's Archaebacteria into a modified archaebacterial group including Halophiles and Methanogens and a few other odd groups and Eocytes, on a lineage close to the Eukaryotes. The Eocytes are mostly thermophilic and sulfur-metabolizing, traits that might have been characteristic of the pre-eukaryotes as well.

DATING THE BRANCHES

WE HAVE thus far omitted assignment of dates or time periods to these three or four ancient branches and to the many small branches that diverge later on the Eukaryotic line during the period of symbiosis. This is because the enterprise of assigning dates to branches of molecular family trees is a risky one and should be separated from the act of constructing the tree. Molecular evidence may be sound for establishing the number of lineages and their branching pattern, but in the billion-year range it offers poor guidance of absolute or even approximate dates. With groups of more recently evolved organisms with clear, well-dated fossil records, it is possible to use the fossil dates to set the time scale of the molecular clock.

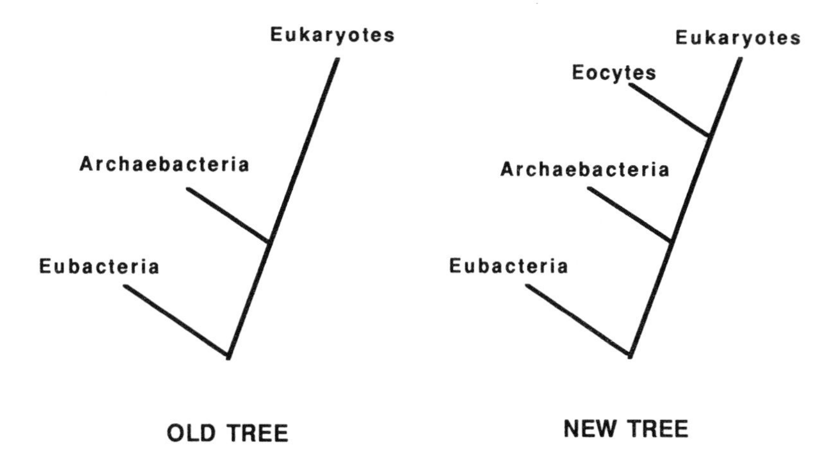

FIGURE 1.5. Rivera and Lake (1992) have proposed that there are four major, deep branches on the family tree of all organisms. After Rivera and Lake 1992.

For example, if we know from paleontological evidence that two groups of mammals diverged five million years ago, we can then count the number of mutations that distinguished specific sequences of their DNA and thus calculate the rate of mutation (per million years, for example). A striking test of this method was carried out by Golenberg et al. 1990, when they isolated chloroplast DNA from 17–20 mya fossils of magnolia and then compared the ancient and modern sequences. A molecular mutation rate derived for lineages in which the fossil record of divergence is well established might then be applied to other related species for which the fossil record is ambiguous, but for which the sequence data can be easily obtained. Thus, on the basis of the sequence, which records the ticking of the molecular clock, the other dates of divergence may be imputed.

This method is used with some confidence on small groups of well-understood organisms with clear fossil records, or even on related genes within a single species. Unfortunately the fossil record of early eukaryotic cells and their predecessors is so sketchy that an absolute rate of mutation could not be assigned to any potential clock molecules. Furthermore, over long periods of time and for large numbers of mutations, the rate of change may be either irregular or difficult to determine. Problems include the fact that just as different molecules might change at different rates (necessitating that highly conserved and slowly changing ones be chosen), different parts of the same molecule might change at different rates depending upon which parts lie in the functional (and more conserved) domains and which do not. This problem is only partly overcome by careful comparison of functionally or structurally distinct parts of sequences.

Another problem with molecular dating techniques, especially over long time periods, turns on the fact that mutation rates may take sudden jumps corresponding to equally sudden cellular events. These might include the insertion or deletion of entire sections of sequence or the duplication of a sequence, leaving the researcher to speculate about what might have been present in the ancestral sequence. A duplication event can have the effect of allowing large numbers of mutations to be fixed in one copy of the sequence, while the other copy maintains some of its original function with a slower rate of change. Indeed some duplicated genes mutate so wildly that they soon become entirely nonfunctional "pseudogenes," but are still retained by the cell.

The rate of molecular evolution is likely also to be affected by the sudden acquisition of symbionts—a profound genetic event for both partners. (This will be discussed in chapter 3 and throughout the remainder of the book.) Many molecules might change their functionality and thus their rate of change under the new circumstances of the symbiosis. If the symbionts arrive in numbers, as they often seem to do, this can have the effect of many gene duplications and an increase in change for some of those sequences. Indeed, symbionts are capable of a level of genetic intimacy that enables the exchange of sequences of DNA, confounding the linearity of the family tree.

All the same, it is tempting to assign dates for early eukaryotic cells, albeit in a careful way (reflecting the cautious interpretations of the fossil record). It is at least possible to look at branching patterns to get an idea of when rapid changes might have occurred and to see whether the fossil record also shows rapid change or commensurate changes in diversity. Andrew Knoll (1992), using data from several molecular evolutionists and many of his own observations of the fossil record, has drawn several conclusions about the correspondence of the molecular family tree of eukaryotic cells with the fossil family tree. The fit between the two trees is surprisingly good, reflecting four bursts of evolutionary activity in the fossil record.

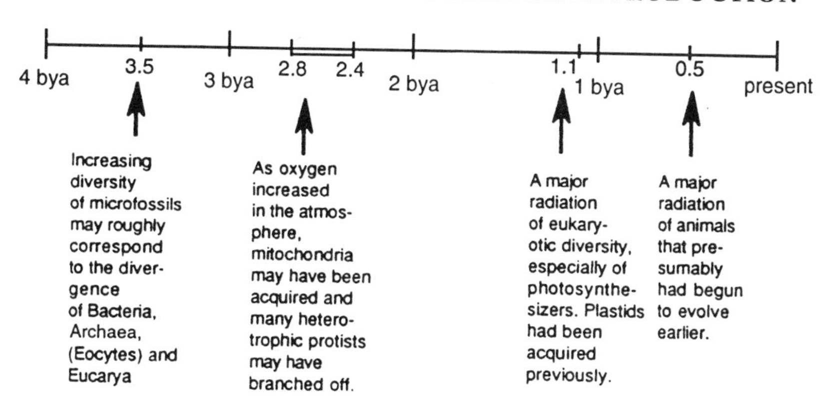

FIGURE 1.6. Knoll (1992) has correlated major events in the fossil record with major branching patterns in molecular evolution. (Scale is billions of years before present.)

First, a period of increasing bacterial diversity seen in microfossils at 3.5 bya may roughly correspond to divergence of the three or four deepest branches: Bacteria, Archea, (Eocytes) and Eucarya. Second, the period of time hypothesized for eukaryote acquisition of mitochondria was extrapolated from the data for the rapid rise of oxygen in the atmosphere (2.8–2.4 bya). This time, too, corresponds to a branching pattern leading to many of the heterotrophic groups of protists. Molecular evidence shows, as well, that some heterotrophic protist groups lacking mitochondria may have branched off somewhat earlier than the rest. Third, a major radiation of eukaryotic diversity, especially of photosynthesizers, is seen in the fossils at 1.1–1.0 bya. The molecular data also suggests that many groups of protists, especially the photosynthetic ones, diversified all at once. Fourth, and finally, Coelomate organisms (fairly advanced animals with distinct body cavities) radiate at 0.56–0.54 bya (having begun their evolution perhaps one hundred million years earlier), yielding an impressive fossil record supported in many parts by molecular archaeology. (See figure 1.6.)

The chronology of these bursts of evolutionary activity will be followed in the next several chapters. Beginning with chapters 2 and 3, the rapid evolution of metabolism in early (prokaryotic) cells will be outlined. In chapters 4 and 5, the branch Eucarya will be discussed in detail as the lineage that gave rise to the pre-eukaryotic host cells that would acquire symbionts and become eukaryotic.

Chapter 6 concerns mitochondria; chapter 7 is about the later-acquired plastids. Chapter 8 describes the evolution of motility organelles. The first eukaryotic motility organelles probably predate both mitochondria and plastids, but we have chosen to put this chapter last, as the evidence is not conclusive and the issue remains in deep controversy. Finally, chapter 9 sets forth an array of problems and questions for future research.

II

The Evolution of Metabolism

MOST OF the major biochemical pathways involving the use of energy and molecules by organisms (metabolism) seem to have evolved with amazing rapidity during the interval between the estimated origin of life 4 billion years ago and the time of the first microfossils 3.5 billion years ago or a little earlier. In a half billion years or less, ancestors of the first organisms to leave a fossil record evolved fermentation and photosynthesis along with several more obscure forms of metabolism. Only respiration and chemoautotrophy seem to have evolved later than the time of the first microfossils, perhaps near the time of the rise of eukaryotes.

A time-traveler to Earth 3 billion years ago would see communities of organisms, some visible to the eye as tufts of green filaments or matlike networks on sediments and in shallow water atop mounds of precipitated minerals and accreted sediments, the by-products of their metabolic activity. These mounds with a veneer of life, called stromatolites, were formed layer by layer over many years; they are a distinctive feature of the early fossil record. Although most stromatolites do not retain clear microfossils, they are evidence of prokaryotic (and, later) eukaryotic metabolism, especially photosynthesis. During their heyday, these underwater structures, like bacterial skyscrapers, could reach meters high. Other prokaryotes 3 billion years ago were forming less visible communities but they too might have been detected as pigmented patches in the soils, sediments, and waters or by the distinctive smells some of them produce, by-products of their metabolism. All of the aqueous bodies and moist soils of Earth must have abounded with prokaryotic communities. Even more unusual environments such as thermal springs at boiling temperatures, salt flats, and deserts were probably habitats for life, as evidenced by extant prokaryotic communities in these environments.

The sequence of events underlying the evolution of diverse metabolisms is a challenge to decipher. Fortunately, four different areas of research each provide evidence, which, in the aggregate, is highly suggestive of a distinctive sequence. Evidence comes from the study of conserved macromolecular sequences such as those of ribosomal RNA, the metabolic pathways themselves, analysis of microfossils, and geochemical signatures of metabolism.

MOLECULAR EVOLUTION OF BACTERIAL METABOLISM

NEW ANALYSES of ribosomal RNA (rRNA) sequences have, during the past few years, brought about a veritable upheaval in bacterial systematics. Bacteria are small and rather uniform in appearance, and therefore morphology is a poor means for distinguishing the major taxa. Instead, the great diversity of bacterial metabolisms, including several types of heterotrophy and autotrophy, have formed the basis for traditional bacterial classification and the organization of such bibles in the field of microbiology as *Bergey's Manual of Determinative Bacteriology,* edited by Buchanan and Gibbons (1974). Indeed many of the practical methods of identification in the microbiology lab have centered around determination of metabolic types.

It was assumed that metabolic groupings were real taxonomic groupings, and perhaps reflected the evolution of bacteria. (One such

TABLE 2.1. The new classification system for bacteria.

Thermatoga
Green nonsulfurs
"Radio resistant" Deinococci and relatives
Spirochetes
Green sulfurs
Bacteroides, flavobacteria, and relatives
Planctomyces and relatives
Chlamydiae
Gram positive bacteria (including *Clostridium* and mycoplasmas)
Cyanobacteria
Purple Bacteria

evolutionary scheme based on metabolic pathways and which still has some validity—in spite of bizarre twists in the rRNA data—is described in a later section.) The entire field of bacterial taxonomy and evolution, however, was apparently due for an overhaul, and it got one, thanks to the persistence and ongoing work of Carl Woese (e.g., Woese 1987) and his colleagues, and others in the field.[1] On the basis of rRNA data, all organisms are now divided into three ancient branches, as discussed in the previous chapter: Eubacteria, Archaebacteria, and Eukaryotes (or Eucarya). (Or, in Lake's classification, Eocytes are added as a fourth branch.) Within the Eubacteria, eleven taxonomic groups are now recognized, many of them bearing little resemblance to the traditional groupings of microbiology. (See table 2.1.) Furthermore, the idea of a linear and easily deciphered series of events in the evolution of bacterial metabolism has been almost entirely overturned. What can be salvaged of the original scheme will be described in a later section of this chapter. First, following the new classification system of Woese (1987), an attempt will be

[1]It may take another full generation of researchers for the new system of classification to be fully accepted and used. Even then, the medical establishment is likely to cling to the old terminology indefinitely. It has certainly done so with respect to fungal taxonomy by using a system so antiquated that it assigns different genus and species names to the same fungus depending on whether it has been isolated from a plant host or an animal host.

made to determine the order of events in the evolution of bacterial metabolism.

Recall from the previous chapter that Eubacteria and Archaebacteria according to Woese are two prokaryotic groups with many extant members. Eucarya is more elusive because the only known extant members (having undergone a transformation by the acquisition of symbionts) are eukaryotes, and as such are not directly useful in determining early prokaryotic evolution. Therefore this analysis concerns the family trees of Bacteria and Archaea. (Archaea is split into Archaea and Eocytes according to Lake, but our discussion here adopts the Woese scheme.)

THE EARLIEST METABOLISM

THERE ARE three early branches on the tree of Eubacteria (figure 2.1) and possibly other early branches as yet undiscovered. On one of the two lowest eubacterial branches is found *Thermotoga maritima,* the only species thus far assigned to this group. *Thermotoga* is anaerobic and extremely thermophilic (growing in environments up to 90°C). It has a fermentative type of metabolism, a simple form of heterotrophy, but it uses some novel enzymes (Juszczak et al. 1991). On the other low eubacterial branch (it is impossible to say which of the two lowest branches came first) are the thermophiles *Aquifex* and *Hydrogenobacter,* which are chemoautotrophs (hydrogen oxidizers) requiring small amounts of oxygen. Chemoautotrophs make their own food (sugar) using energy from an inorganic molecule, in this case, molecular hydrogen. However, hydrogen oxidizers are usually capable of heterotrophy as well.

As with Eubacteria, the Archaebacteria family tree also supports two branches at its base (figure 2.2). One branch comprises a fairly uniform group of extreme thermophiles, most of which are anaerobic and use sulfur compounds as a source of energy. The other branch has at its base a thermophile, although the rest of the group includes members that grow at lower temperatures. However, many of these low-temperature members also have exceptional habitats and metabolism, including halophily (love of extreme salt) and methanogenesis (production of methane). The thermophiles at the base of the Archaebacteria are anaerobic and some have in common an ability to carry out fermentation.

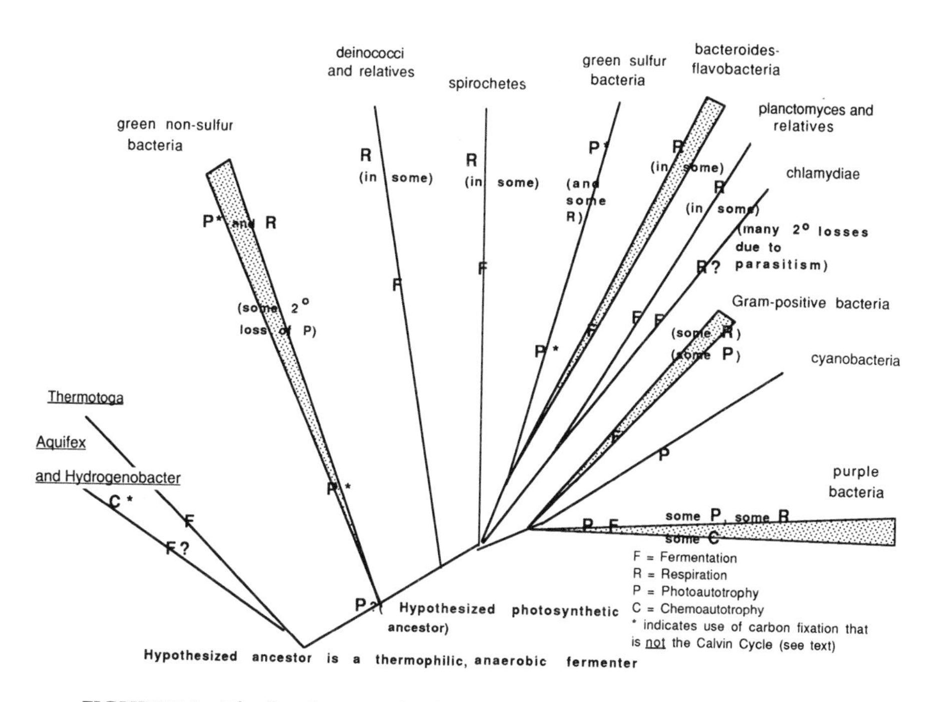

FIGURE 2.1. The family tree of Eubacteria. After Woese 1987.

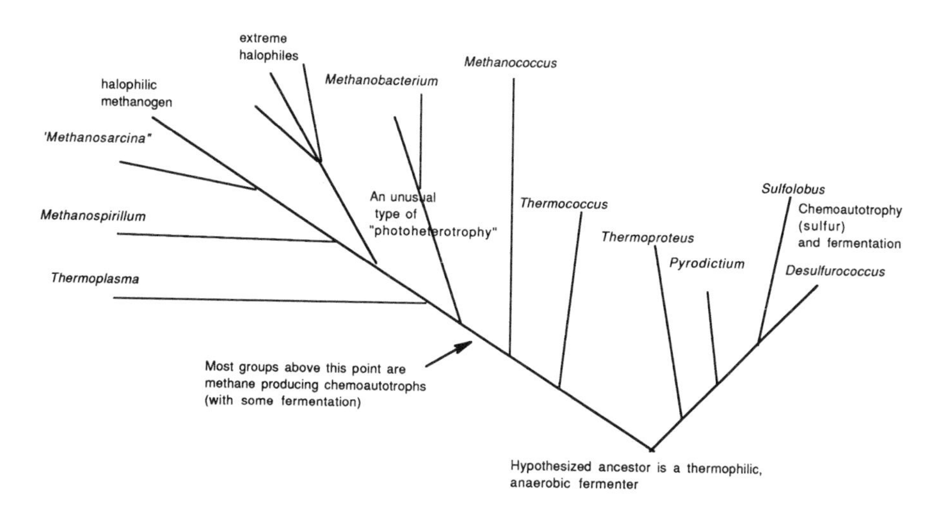

FIGURE 2.2. The family tree of Archaebacteria. After Woese 1987.

Thus at least three characteristics are common to the base of both Eubacteria and Archaebacteria: an ability to ferment (considered a fairly simple form of metabolism), thermophily, and anaerobiosis. The early evolution of these three features in the Woese system of classification is consistent with other evolutionary schemes and even with some hypotheses concerning the origin of life.[2] Notably, there is little doubt that the early atmosphere was lacking in oxygen, thus the earliest organisms would have been anaerobic. Small fermentable food molecules might have been available to be scavenged, having been produced by the same chemical reactions that created the "primordial soup" out of which life arose. It is also widely held that the surface of Earth was a rather hot place during its formation and early years. At the time of the origin of life (about 4 bya) hot springs were probably abundant. What happened next to these anaerobic, fermenting, hot spring dwellers must have been a rapid and multibranching radiation such that in a short period of time (perhaps a mere half billion years) most bacterial metabolisms had evolved from fermentation, including some fairly exotic variations.

PHOTOSYNTHESIS EVOLVES

NEXT IN antiquity on the tree of Eubacteria are the Green Nonsulfurs, filamentous gliders some of which, such as *Chloroflexus*, perform a strange sort of photosynthesis. *Chloroflexus* synthesizes sugar by using small organic molecules both as a source of carbon (rather than carbon dioxide) and as a source of hydrogen (rather than hydrogen sulfide or water). The pathway for synthesizing the sugars present in *Chloroflexus* is an unusual one and appears to be part of what would become the Krebs cycle, used by aerobic heterotrophs to

[2]This brief description is not intended to be a full review of all possible interpretations and hypotheses drawn from the scanty data on early metabolism. Nor do we endorse with confidence the hypothesis that fermentation came first. An important alternative hypothesis is that chemoautotrophy (perhaps of a hydrogen oxidizing type) was at the base of the prokaryotic tree. It depends in part on how to interpret the extant hydrogen oxidizers *Aquifex* and *Hydrogenobacter*, which are both heterotrophic and autotrophic (e.g., Wächershauser 1990; Kandler 1993). It seems reasonable to us, however, that, because those hydrogen oxidizers do require oxygen, this aspect of the metabolism may have evolved later as trace oxygen finally became available. We thus suspect that heterotrophy is actually at the base of that lineage.

break down food molecules. However, in this case that part of the cycle is being run "backwards" to synthesize molecules.

Photosynthetic pigments must have had their origins in green nonsulfurs. This branch is also the likely progenitor of the electron transport pathway needed to process energy-rich molecules. The use by photosynthetic nonsulfurs of small organic molecules to make larger ones puts their metabolism rather close to heterotrophy, and indeed heterotrophy is one of their options.

Extant green nonsulfurs are actually quite versatile. *Chloroflexus* is able to use sulfur compounds as a source of reducing power, as can some photosynthetic groups farther up on the family tree. *Chloroflexus* is even capable of a type of aerobic heterotrophic growth, presumably derived from fermentation. Secondary loss of photosynthesis (but retention of some pigments) characterizes a close relative of *Chloroflexus, Herpetosiphon,* as a heterotrophic aerobe. The fact that *Herpetosiphon* is aerobic suggests that the secondary loss occurred when oxygen had accumulated in the atmosphere. Furthermore, it suggests that a Krebs cycle or Krebs cyclelike pathway and an electron transport system (necessary for aerobic heterotrophy) must have evolved fairly easily from photosynthetic metabolism considering that a reversed part of the Krebs cycle was apparently already being used for the synthesis of sugars. Easily, because this kind of electron transport mechanism apparently occurred several times in the other lines of Eubacteria.

A DIVERGENCE OF METABOLISMS

AFTER THE green nonsulfurs branch off, nine other groups of Eubacteria diverge almost all at once, and so the sequence of events is difficult to determine. This radiation can be explained by invoking numerous modifications and secondary losses of metabolic capabilities already developed. The branch Purple Bacteria is an excellent example (table 2.2). Of four groups of purple bacteria—α, β, γ, and δ—three have members that include photosynthesizers, chemoautotrophs, and heterotrophs. (These chemoautotrophs synthesize sugars out of carbon dioxide, usually using a sulfide for reducing power and energy.) Some of the branches of purple bacteria include aerobes, which must have evolved later, and many times independently, when oxygen accumulated in the atmosphere.

TABLE 2.2. The diverse Purple Bacteria.

α subdivision
Purple nonsulfur bacteria, rhizobacteria, agrobacteria, rickettsiae, *Nitrobacter*
β subdivision
Rhodocyclus, (some) *Thiobacillus, Alcaligenes, Spirillum, Nitrosovibrio*
γ subdivision
Enterics, fluorescent pseudomonads, purple sulfur bacteria, *Legionella,* (some) *Beggiatoa*
δ subdivision
Sulfur and sulfate reducers *(Desulfovibrio),* myxobacteria, bdellovibrios

Woese hypothesizes that at the base of the purple bacteria is a photosynthesizer with purple pigments; thus the name of the group, even though it contains many nonpigmented members (Woese 1987). The heterotrophy characteristic of this group is explained as secondary losses of photosynthesis. Two alternative hypotheses are rejected (Woese et al. 1980). One is that photosynthesis based on carbon dioxide and water, a complex process with many enzymes, somehow arose independently several times. The other is that the photosynthetic genotype was transferred across groups. This too is unlikely in that it would require that many genes be transferred en masse in exactly the same way, even though the mechanisms of transfer as they are understood today are too random for this to be conceivable (Woese et al. 1980). (Chapter 3 explores this point in detail.)

If photoautotrophs with purple pigments are at the base of the purple bacteria, then the three types of chemoautotrophy in this group must have arisen independently. These three subgroups are the Nitrobacters, the Thiobacilli, and the Beggiatoans. Their independent evolution is not intuitively obvious but can be explained in that each type probably evolved in different environments and therefore had different substrate molecules available to them as sources of energy. These chemoautotrophs have in common the nearly universal enzyme pathway called the Calvin Cycle (used to make sugars), which all three would have inherited from a photosynthetic ancestor. The Nitrobacter group uses ammonia or nitrite as a source of energy. The Thiobacillus group uses iron or sulfur compounds as a source of energy, and the Beggiatoans use sulfur compounds. All three groups need at least trace amounts of oxygen to efficiently accomplish the oxidation of the nitrogen, iron, or sulfur substrates. However, too much oxygen would convert these substrates to unusable oxidized forms. Thus many of these chemoautotrophs, which now thrive in aerobic-anaerobic transition zones in sediments and waters, might have been especially successful at a time when oxygen was beginning to accumulate in the atmosphere. Many of the extant members of these chemoautotrophic subgroups also retain heterotrophic capabilities, reminders of their heterotrophic ancestry.

A SUMMARY OF TRENDS IN EUBACTERIA

IN SUMMARY, metabolic evolution within the Eubacteria branch apparently begins with thermophilic, anaerobic fermenters represented by extant *Thermatoga.* The lineage that contains *Thermatoga* then gave rise to an obscure group of photosynthesizers, green nonsulfur bacteria (e.g., *Chloroflexus aurantiacus,* a thermophile). Assuming that photosynthesizers are at the base of all the other nine groups, considerable secondary loss must be invoked to explain the diverse forms of heterotrophy. Five of these nine groups—Spirochetes, Bacterioles, Planctomyces, Chlamydiae (which may have lost many capabilities on becoming obligate pathogens), and Radioresistant—seem to have lost photosynthesis in all members, reverting to fermentation (a capability never entirely lost from the ancestral group). Some subgroups of these five types of nonphotosynthesizers became aerobic heterotrophs, presumably when oxygen had accumulated. The pathways of photosynthesis must therefore be the foundation for respiration, especially pigment synthesis, and electron transport.

The other four of the nine later branches of Eubacteria evolved very successful variations of photosynthesis. Green sulfur bacteria and cyanobacteria each form their own cohesive group in which photosynthesis is the exclusive form of metabolism. Cyanobacteria with their oxygen-generating photosynthesis are perhaps the most important organisms in the history of the Earth. They evolved the type of photosynthesis that dominates the food web today, and indeed some cyanobacterial lineages continue as major participants in the guise of plastids of algae and plants. Furthermore, oxygenic photosynthesis is almost solely responsible for the accumulation of oxygen in the atmosphere, which had such a profound effect on all subsequent evolution.

Another group, the gram-positives, consists mostly of heterotrophs but also includes the photosynthetic heliobacteria. And finally, the purple bacteria (previously discussed), evolved their own unique styles of photosynthesis. Three new types of photosynthesis may have come somewhat later and independently in the purples after many heterotrophic branches had lost the capability to perform the photosynthesis of their ancestors. As in the previously discussed branches, a reversion to fermentation (in some lines) gives rise to respiration. Furthermore, at some point when sufficient oxygen became available, chemoautotrophy arose in three, different lines.

The evolutionary picture is a messy one, but its visual presentation in figure 2.1 is our attempt to make sense of it all. No attempt was made to adjust the diagram to scale except that some of the derived metabolisms, respiration and chemoautotrophy, must have evolved after photosynthesizers generated oxygen, and we have therefore provided for such in the illustration.

EVOLUTION IN THE ARCHAEBACTERIA

BASED ON the work of Carl Woese (1987), a family tree for the other main group of prokaryotes, the Archaebacteria has been constructed (see figure 2.2). As with the Eubacteria, the progenitor of the Archaebacteria was likely a thermophilic, fermenting anaerobe. However, members of the Archaebacteria appear not to have diverged as far from the progenitor as have the Eubacteria. On the other hand, a lack of information about the metabolism in many of these Archaebacteria may present a picture of false simplicity.

It appears that several varieties of fermentation evolved in the Archaebacteria. Some of these independently gave rise to at least two types of chemoautotrophy: methane-producing and sulfur types, which use two distinct and unusual pathways that resemble parts of the heterotrophic Krebs cycle but run backwards. Some members of the methane-producing line apparently lost the chemoautrophic ability and became fermenters. One strange shoot of that branch, the extreme halophiles, has evolved a unique photoheterotrophic type of metabolism.

SEMES AND THE EVOLUTION OF BACTERIAL METABOLISM

SADLY, THERE is not much left of the linear, easy-to-teach, easy-to-learn system for bacterial phylogeny now that the tree derived from the rRNA data has made it obsolete. However, the core logic of the old system does mesh with the new rRNA data, albeit in a complicated way (Margulis 1982). The logic is this: bacterial metabolic pathways may be arranged approximately from simple to complex, and the pathways are composed of simple units that tend to be conserved in evolution. In fact, these units are really not so simple, and therein lies the key to their conservation; even the simplest units of metabolism are the products of more than one gene—they are multigenic traits or "semes" and are analogous to building blocks.

If one gene in a seme mutates such that the function of the entire seme is lost, the organism may be said to have "lost" the function, however it very likely maintains most or even all of the genes that made up that trait. This phenomenon has been called "crypticity" by Ripley and Anilionis (1978), who consider it to be one of the major means by which bacterial evolution occurs. Such "cryptic" genes may continue to mutate, some may be lost (e.g., through errors in replication), but there is also the possibility that the entire seme may be "revived" at a later point. In effect, the building block itself was not lost but merely tossed back into the toy box temporarily. If many mutations have occurred in the interim, the new function may be quite different from that of the ancestral building block, but the complement of genes making up the seme may still be recognizable. Thus one can, theoretically, trace the history of bacterial metabolism by tracing the history of the relevant semes—their evolution, their combinations with each other, their secondary "losses," and their "revivals" in new forms and functions.

How and why are the genes responsible for a trait maintained in the absence of positive functional selection? The answer is not clear, but may be found in the fact that many proteins are known to have multiple functions, and sometimes a single protein functions in disparate biochemical pathways. A simple example is the ability of an enzyme to catalyze both the forward and reverse directions of a single reaction, as in the ATP-using or -making proton pump described earlier.

Indeed, in evolutionary terms, the "function" of an enzyme is merely the major function we associate with it at a given point in time—a "snapshot" of the enzyme on its way along a continuum of functions through time. The same is presumably true of well-coordinated multigenic traits and semes.

If the amino acid sequence of an enzyme was under selective pressure from multiple functions, it might be conserved enough to allow the reclamation of a primitive function after many generations of inactivity. In addition, "doubling-up" of functions may have been an important mechanism in ancient organisms with smaller genomes than those of today's bacteria.

For this discussion, we will concentrate on Eubacteria since much is known about that group's metabolism. Anaerobic fermentation is widely regarded as one of the most basic and ancestral semes. In a few bacterial groups it may have persisted through to today (for example, in deep anaerobic mud) in something near to its original form, but in most groups either the ancestral form or some variation of anaerobic fermentation seems to be a revival of a seme function once lost, although the genes were maintained. Fermentation may be simplified and generalized to a scheme that looks like this:

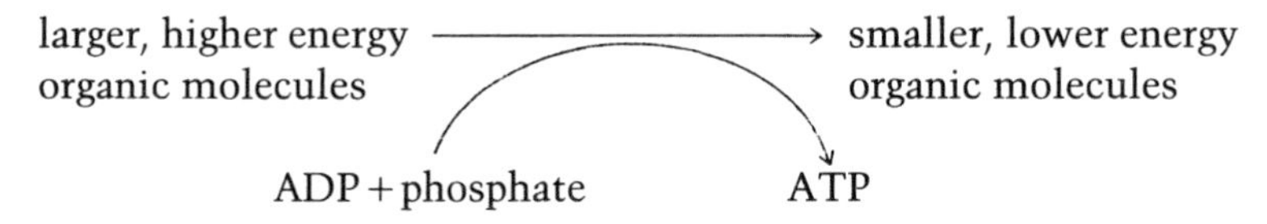

Fermenting bacteria can obtain and store energy by breaking down chemical bonds in a high-energy molecule such as a sugar and then using the captured energy to build molecules of adenosine triphosphate (ATP). Most of the useful energy of ATP is stored in the final phosphodiester bond added when the third phosphate is attached. The waste product of fermentation is a somewhat smaller, less energy-rich molecule, such as an acid or a gas like carbon dioxide or molecular hydrogen.

The most ancient but still-extant example of a fermenter is probably *Thermatoga,* the marine, thermophilic fermenter, that occupies an early branch of the Eubacteria. However, *Thermatoga* likely has other characteristics that are not so ancient, in that it cannot have stood absolutely still evolutionarily for more than three billion years. (See Juszczak et al. 1991 for a summary of this fermenter's unique characteristics.) Nevertheless, *Thermatoga* offers major clues about what the earliest semes might have been.

All other fermenters on the Eubacterial tree appear to have secondarily lost the function and then regained it later (thanks to not having actually lost the genes for it). *Thermatoga* takes up molecules such as sugars and other carbohydrates and gives off as waste lactate, acetate, carbon dioxide, and molecular hydrogen. *Thermatoga* also has an enzyme complex, called a hydrogenase, containing iron and sulfur, and it has a ferredoxin (both of which transfer electrons). Although the function of these molecules is not well understood in *Thermatoga,* the semes that produce these complex molecules may be universal in the Eubacteria. For example, in the more recent metabolisms of various kinds of photosynthesizing and respiring bacteria, iron-sulfur complexes participate along with cytochromes and, in some cases, ferredoxins in electron transport chains for respiration and photosynthesis.

The situation is similar for *Aquifex* and *Hydrogenobacter,* which are extant representatives of an ancient group of chemoautotrophs that can also be heterotrophic. Which came first, the heterotrophy or the chemoautotrophy, is unsettled. However, we will assume that chemoautotrophy came later, as it is more complex and requires a measure of free oxygen that would not have been present in the early atmosphere.

PHOTOSYNTHESIS AS A SEME

FERMENTATION IS a somewhat inefficient metabolism in that its waste products include molecules that are themselves food for other forms of metabolism, such as respiration. Furthermore, fermentation depends upon the existence of food molecules (large, energy-rich organic compounds) in the environment, a situation that might have existed at the origin of life. However, after the evolution of fermentation, most food molecules would have been scavenged, leaving fermenters in something of a "crisis" (Margulis 1982).

Thus it is not surprising that the evolving photosynthesizers might have been favored shortly after the evolution of fermenters, as they are able to make their own food using energy from the sun along with some source of carbon and hydrogen. Some obligate photosynthesizers lost the function of fermentation, but probably not the genes for it—making possible revivals of fermentation by later Eubacteria. Why would a wasteful form of metabolism suddenly be advantageous again? Presumably after the spread of photosynthesis increased the biomass of the planet, an abundance of food molecules would have been available again not only for photosynthesizers themselves to exploit for nutrient value but for various heterotrophs, who would obtain both energy and chemical building blocks from these food molecules. Some photosynthesizers are indeed facultative; they can use heterotrophic pathways when appropriate food molecules are available.

Those first photosynthesizers are represented today in an obscure group: the Green Nonsulfurs. This group is the second oldest branch in Woese's scheme. An extant example is *Chloroflexus*—a filamentous, gliding bacterium, which lives in thermal springs (Castenholz and Pierson 1981). Chloroflexus makes food molecules (carbohydrates) by using energy from the sun. But curiously, it uses smaller food molecules (rather than carbon dioxide and water or hydrogen sulfide) as sources of carbon and hydrogen. Thus its form of metabolism would fit well with its fermenting predecessors in that it would use the waste products of fermentation.

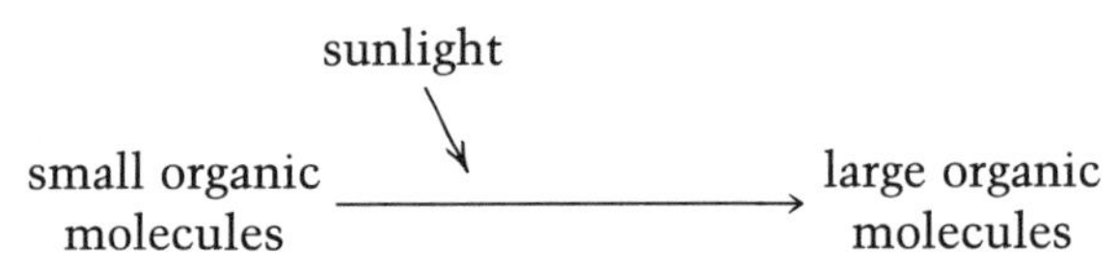

Chloroflexus can also perform a more complete photosynthesis, starting with carbon dioxide and molecular hydrogen. In addition *Chloroflexus,* which is normally an anaerobe, can switch to a type of aerobic respiration that probably evolved from the fermentation seme it retained from its ancestors. Thus, versatile *Chloroflexus* is an organism at a crossroads. It seems to have retained the fermentation pathway of its ancestors, as well as evolved a partial form of photosynthesis similar in some ways to fermentation but also capable of using carbon dioxide and molecular hydrogen. Furthermore, at some later point it evolved respiration, but did not discard any of the previous capabilities. *Chloroflexus* is thus a collector of semes with an unusually full attic of semes that still function. One of its relatives, *Herpetosiphon,* in contrast, lost its photosynthetic capabilities while retaining some photosynthetic pigments.

Photosynthesis forms the base of the next junction on the Eubacterial tree, where nine branches diverge virtually simultaneously. These nine branches display both of the two possibilities for evolutionary trends emanating from a photosynthetic ancestor: descendants can evolve different types of photosynthesis through modifications of the original photosynthetic seme or they can lose photosynthetic capabilities (though perhaps not all of the genes) and revert to fermentation—one of the most conserved and nearly ubiquitous semes on the family tree. Innovations in the photosynthetic seme mostly involve changes in pigments and the use of different molecules as donors for hydrogen. For example, hydrogen sulfide is used for hydrogen by the photosynthetic green sulfur and purple sulfur bacteria; alternatively, water is the source of hydrogen for cyanobacteria, who do an "oxygenic" form of photosynthesis.

CARBON FIXATION AS ONE OR MORE SEMES

AT ONE time carbon fixation (the synthesis of sugars) was understood to be a single, though complex cycle—the Calvin cycle (discussion and figure in chapter 6)—and typical of all of the most common and visible photosynthesizers: plants, algae, and cyanobacteria. The picture is no longer so simple, now that several alternatives to the Calvin cycle are known.

Recall that *Chloroflexus* employs a (presumably ancient) form of carbon fixation by taking part of the heterotrophic Krebs cycle (description and figure in chapter 7) and running it backwards. The chemoautotrophs *Aquifex* and *Hydrogenobacter* fix carbon with another kind of reversal of the Krebs cycle, as do some of the sulfur-metabolizing Archaebacteria. Other groups of Eubacteria and Archaebacteria use yet another modified (backwards) version of the Krebs cycle. Even the Calvin cycle itself, characteristic of all eukaryotic photosynthesizers as well as cyanobacteria and autotrophic varieties of purple bacteria looks like part of a heterotrophic pathway in reverse.

Do these clues indicate that aspects of heterotrophy are ancestral to carbon fixation, rather than the reverse? We favor the hypothesis that heterotrophy is ancestral because fermentation is the simplest known form of metabolism and because bacteria at the base of the family tree have in common some form of heterotrophy (albeit sometimes accompanied by a diversity of other metabolic pathways, which we believe must have evolved later).

ELECTRON TRANSPORT AS A SEME

ELECTRON TRANSPORT pathways as mechanisms for storing chemical energy also diversified early in the history of life. Fundamental to these pathways is the molecule adenosine triphosphate (ATP), whose high-energy bonds can be used for energy storage and retrieval on demand.

Two components of many electron transport systems, ferredoxin and iron-sulfur proteins, are found in *Thermatoga,* although they do not seem to be part of an electron transport system. Their function in *Thermatoga* is unknown. In *Chloroflexus,* which appears in a later branch of the ancestral line that contains *Thermatoga,* electron transport functions as part of the so-called green photosynthetic system. What happened in the components of the family tree between that of *Thermatoga* and *Chloroflexus?*

One hypothesis has to do with the way some electron transport molecules function in extant fermenting bacteria. Fermenters that excrete acids (as do *Thermatoga* and many others) create an acidic environment high in positive hydrogen ions, H^+. H^+ ions have a tendency to leak back into the cell which must continually pump them out in order to maintain the correct pH—the correct level of H^+ ions. Some of these pumps use the energy stored in ATP. Other pumps use electron carriers embedded in the outer membranes, which remove electrons from the acidic waste molecule and in doing so "pump" H^+ across the membrane without using ATP. These electron transport pumps can be so efficient that a high concentration of H^+ builds up outside of the cell membrane. But the excess H^+ ions also leak back through the ATP pumps, essentially causing them to operate in reverse. An ATP pump operated in reverse does not use ATP but rather makes it. The major innovation of the green photosynthesizers and their descendants was to take the established system of ATP pumps and electron transport pumps (now called the electron transport pathway) and drive the system in reverse with some cheap, available form of energy, such as sunlight (see figures 2.3 and 2.4).

Sunlight can stimulate many chemical reactions, especially in pigmented molecules that absorb particular wavelengths of light. One such chemical reaction occurs in the light-absorbing molecule chlorophyll. When a chlorophyll molecule is exposed to sunlight, an electron breaks free and thereby initiates a chain of events along the

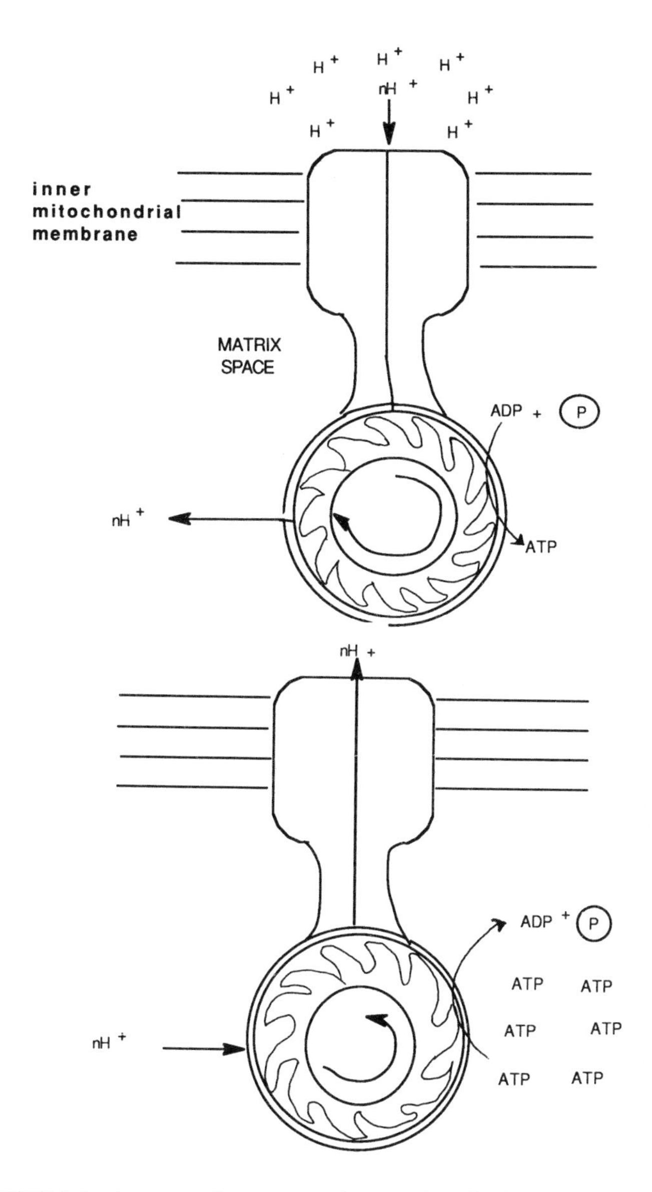

FIGURE 2.3. A proposed sequence of events by which ATP-requiring proton (H^+) pumps evolve into ATP-producing proton channels. In the first picture the ATPase operates such that H^+ pass into the matrix space and cause ATP to be formed. In the second picture H^+ moves in the opposite direction and is essentially pumped out while ATPs are used up. After Alberts et al. 1983.

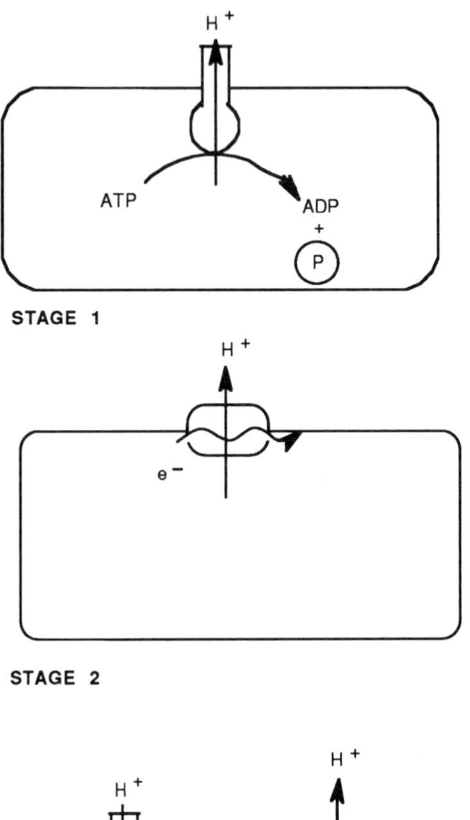

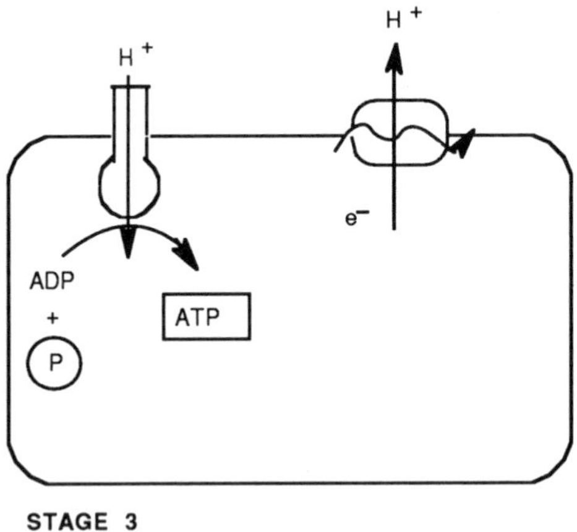

FIGURE 2.4. A hypothesized evolutionary sequence: In Stage 1 the cell pumps out excess H^+ using up ATP. In Stage 2 the cell uses an electron transport system to pump out H^+ and thus does not waste ATPs. In Stage 3 the two systems are combined, except now the efficiency of the H^+ pump enables the ATPase to be run backward to make ATP. After Alberts et al. 1983.

electron transport pathway, such that positive hydrogen ions are moved across a membrane and put to work building the high-energy bonds of ATP. Chlorophyll-mediated photosynthesis was an important innovation in metabolism not only because of its energy capture and storage ability but also because its mode of electron transport, as a seme, was then available for modification into other forms of metabolism, such as aerobic respiration (Alberts et al. 1983).

PHOTOSYNTHETIC PIGMENTS AS SEMES

PHOTOSYNTHETIC PIGMENTS constitute the other major seme (in addition to the seme of electron transport) that evolved in the interim between the origin of ancestors to *Thermatoga* and the ancestors of *Chloroflexus.* The evolution of the major photosynthetic pigment, chlorophyll, will be the focus here although other pigments of secondary importance such as carotenoids also evolved during this time.

The ancestral molecule (also a seme) that gave rise to chlorophyll and to many other important molecules of the cell is the highly conserved porphyrin ring (figure 2.5). Some porphyrin derivatives that are pigmented electron carriers, the cytochromes, are nearly ubiquitous, as they are part of all known electron transport chains. Variations on porphyrin rings yield green pigmented chlorophylls that react to certain wavelengths of light by releasing electrons to an electron transport pathway. If cytochromes were early components of a mechanism to pump H^+ ions out of fermenting cells, then some of these complex molecules might have evolved to make the earliest chlorophyll, thus converting the ion pump system to photosynthesis. Porphyrin-based molecules have not, however, been found in *Thermatoga,* suggesting an evolution during the *Thermatoga-Chloroflexus* interim. Nevertheless, lack of porphyrin-based molecules (rare as it is) cannot always be construed to be a primitive condition.

Parasites can be deceptive because the rich environment of a host organism allows many secondary losses of function and even genes to occur. Parasites may appear to be simple (and thus primitive) but the condition is usually highly derived and the consequence of considerable loss of redundant functions. *Clostridium,* a member of Eubacteria that is an obligate anaerobic parasite, is an important

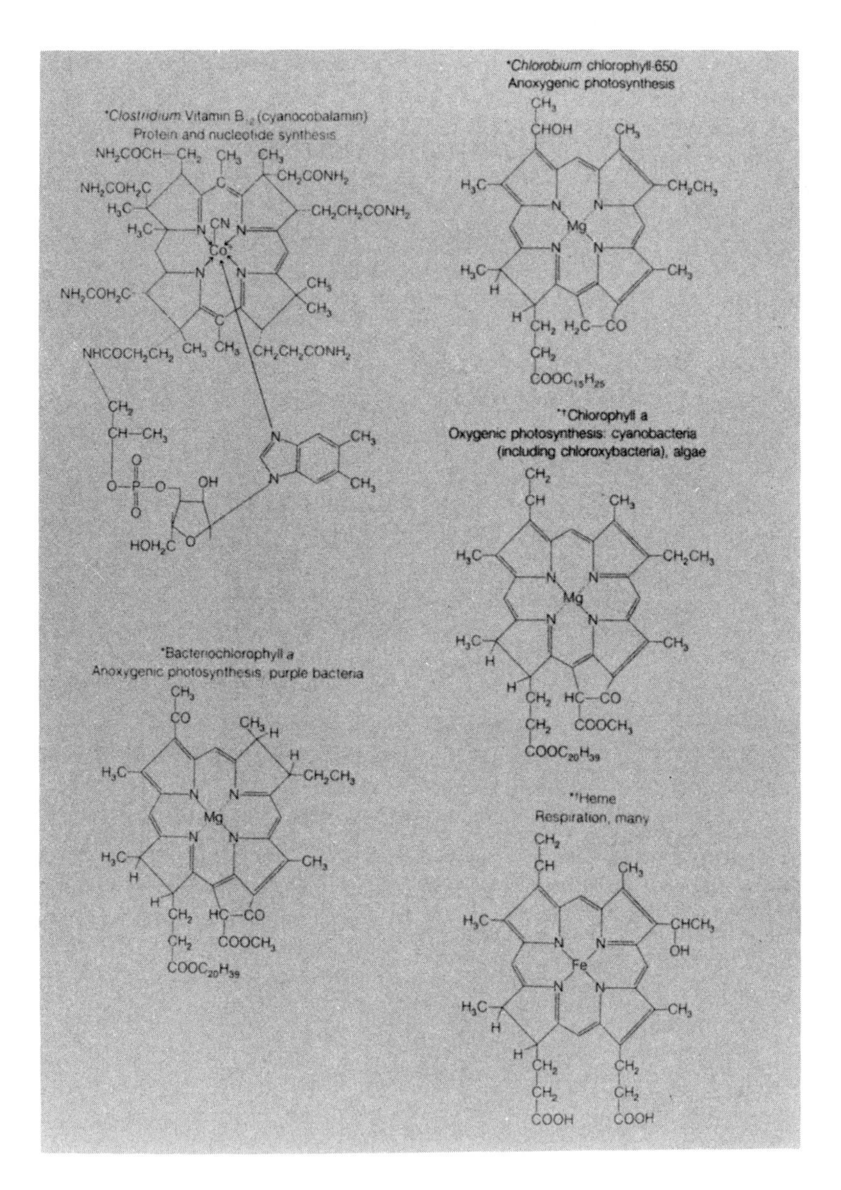

FIGURE 2.5. The versatile porphyrin ring gave rise to various cytochromes, hemoglobins, and chlorophylls. Reprinted from Margulis 1993.
* present in Eubacteria
† present in Eubacteria and eukaryotes

example because it was long considered to be a primitive form of fermenter, lacking even cytochromes (Margulis 1982). However, *Clostridium* is now known on the basis of rRNA data (Woese 1987) to be a fairly advanced member of the Gram Positive group of Eubacteria, and must have lost its ability to make porphyrin ring molecules (a function now taken up by its obligate hosts). *Clostridium*, like most Eubacteria, descends from a photosynthetic ancestor.

VARIATIONS ON THE FERMENTATION SEME

VARIATIONS ON fermentation occurred in many of the other lines of Eubacteria, those that were not devoting themselves to photosynthesis. A photosynthetic world is one that generally supplies an abundance of food, usually in the form of wastes and dead cells, available for use as substrates by the fermenters. The evolving photosynthetic world on the early earth was a good place for fermenters to evolve too. The "food crisis" would have been over. Thus fermentation, carried along as a seme from the days of *Thermatoga*, appears again and again in Eubacteria.

Innovations included the use of electron transport to generate more and more molecules of ATP from the same amount of food. In the highly inventive Purple Bacteria group, the extant *Desulfovibrio* uses electron transport as a means of squeezing additional energy out of what, in related species, are simply discarded waste products of fermentation. *Desulfovibrio* sends electrons from waste organic acids through an electron transport chain to make more ATP. At the time of its origin, this pathway is likely to have been similar or even identical to the electron transport chain that most fermenters would have been using to get rid of excess hydrogen ions.

Desulfovibrio added a new part to electron transport; it found a way to keep the electrons flowing continuously (like a river) by removing them at the end of the line with an "electron acceptor." Otherwise a backlog of electrons at the final electron transport carrier would have the effect of damming the stream of electrons and preventing further flow. (Other organisms such as cyanobacteria solved this problem with a cyclic flow of electrons.) *Desulfovibrio* uses sulfate compounds available in its environment to react with the electrons and some of the H^+ to form hydrogen sulfide, a volatile waste product that gives marine and estuarine sediments in which

Desulfovibrio are active the smell of rotten eggs. Sulfates are used in a way that is analogous to the way oxygen is used in respiration. Thus *Desulfovibrio* may be thought of as a sulfur respirer.

The Calvin cycle, the method of carbon fixation for many photosynthesizers on the later branches of the family tree, may be regarded in part as a reverse of fermentation. Furthermore, other, obscure types of carbon fixation (discussed in a previous section) may be thought of as variations (reversals) of parts of heterotrophic reactions. Thus, although our conclusion is admittedly speculative, the carbon fixation aspects of autotrophy may be thought of as semes of some part of fermentation or other heterotrophic processes.

OXYGEN RESPIRATION AS A SEME

THE JUXTAPOSITION of fermentation and electron transport occurred so many times independently in the Eubacteria that it must have been an extraordinarily easy step to link the two. Respiration (using oxygen), which evolved independently again and again in Eubacteria, is an efficient form of metabolism involving fermentation steps, followed by electron transport (to generate abundant ATPs) and finally by the use of free oxygen as a final electron acceptor. The waste product is innocuous: water.

Oxygen accumulation in the atmosphere provided the major selection pressure for respiration, an event that would not have occurred until the cyanobacterial form of photosynthesis had been established. Early respiration might have been more of an oxygen detoxification mechanism, one of many that evolved to deal with a gas that is both toxic and corrosive. Many of the detoxification mechanisms extant today involve porphyrin-ring molecules, such as the cytochrome positioned at the end of electron transport to capture the oxygen. Combining an efficient form of metabolism (respiration) with oxygen detoxification (otherwise an expensive process) must have been wildly successful in many groups, especially as oxygen continued to accumulate. Those groups that did not invent ways to cope with or profit from oxygen went underground or underwater into anaerobic zones, or took refuge in the relatively anaerobic cytoplasms of hosts as some of the first symbionts.

CHEMOAUTOTROPHIC USE OF SEMES

CHEMOAUTOTROPHY IN the Eubacteria was apparently not so easily evolved in most groups, with the exception of the ancient lineage of *Aquifex* and *Hydrogenobacter* and the more derived Purple Bacteria. Chemoautotrophy in the purple bacteria utilizes most of the typical photosynthetic semes, the Calvin Cycle, and electron transport—but relies on reduced minerals as a source of energy and hydrogen ions rather than sunlight and water or hydrogen sulfide. (Specifically, nitrobacteria use ammonia or nitrate and Thiobacilli and Beggiatoans use sulfur compounds.) One might conclude that something about the way in which photosynthesis evolved in the purple bacteria was especially conducive to the evolution of chemoautotrophy, as it happened three times independently in the Purples but in no other advanced Eubacteria.[3]

One genus of purple bacteria that is a sulfur photosynthesizer, *Thiocapsa,* can, in darkness, switch to a sulfur-based chemoautotrophy. This ability makes the ancestors of *Thiocapsa* a possible evolutionary link between the two metabolisms.Two of the chemoautotrophic Purple Bacteria use reduced sulfur or iron compounds, which would be prevalent in the preferred habitat of purple bacteria, sulfureta (sulfur-rich environments). As trace amounts of oxygen became available, conditions may well have been conducive for the evolution of chemoautotrophs.

SEMES AS BUILDING BLOCKS

IN SUMMARY, the rRNA data has considerably complicated but also enhanced the picture of metabolic evolution by forcing a recognition that semes (genetic building blocks) may lose their function, their phenotypic expression, but are seldom completely lost as genetic

[3]The fact that the chemoautotrophs are on relatively recent branches of the Eubacterial family tree means that those few food webs with chemoautotrophy at their base, such as the deep sea hydrothermal vents, may not be models for ancient community structures, but rather for later developments of higher complexity. Many modern chemoautotrophs also retain the function of the ancestral seme, enabling them to ferment under certain conditions. Even the most ancient branch of chemautotrophs requires oxyen, and many other branches evolved that type of metabolism later.

units. As such, they may be revived or modified and revived in some new function (table 2.3). The old picture was an unnaturally linear progression:

fermentation → sulfur breathing → photosynthesis → chemoautrophy or respiration

It wrongly assumed entire losses of some genetic semes whenever the phenotypic expression was lost, and it forced too many disparate organisms into the same taxa based on metabolism alone. A conclusion that underlying genetic diversity was retained even when the phenotypic expression of such diversity was lost is more in keeping with the incredible versatility and flexibility displayed by most bacteria.

ISOTOPIC EVIDENCE OF METABOLISM IN THE FOSSIL RECORD

THE HIGHLY branched sequence of events in the evolution of metabolism that was established in the first part of this chapter is based mostly upon rRNA sequence analysis and semes (multigenic traits). Here the geological data will be presented and will, in part, be used

TABLE 2.3. Expression of semes in metabolic functions.

The Metabolisms (not necessarily in order)	*Fermentation*	*Electron Transport*	*Photosynthetic Pigments*	*Carbon Fixation* *	*Krebs Cycle*
Fermentation	√	some			
Photosynthesis		√	√	√	
Sulfur breathing	√	√			
Oxygen breathing (respiration)	√	√			√
Chemoautotrophy		√		√	

NOTE: a check mark in the box indicates an active seme *but* lack of a check mark means only loss or lack of function, not necessarily loss or lack of the genes comprising the seme. Therein lies the amazing versatility of evolving bacteria. (*Thermatoga* may be one of the few bacteria to completely lack most of the semes although even if it may be found to have some of the rudiments.)

* There may be at least four independent variations of carbon fixation, all (we believe) derived from heterotrophic semes.

to root the evolutionary tree and to establish dates and time intervals during which the evolution of metabolism occurred. The information comes from two major sources: the ratios of certain isotopes preserved in the rocks and the microfossils and trace fossils of early organisms. As the molecular phylogenies have indicated, the evolution of metabolic pathways was surprisingly rapid, and this is confirmed by the fossil evidence described next.

It happens that many elements are mixtures of isotopes; that is, there are variations in the number of neutrons that accompany a precise number of protons in the nucleus of an atom. Mixtures of heavy and light isotopes deposited in sediments and trapped in the shells and other hard parts of organisms are sometimes preserved in the fossil record in their original ratios. It also happens that organisms show preferences in metabolism for particular isotopes; they tend to fractionate isotopic mixtures, leaving signatures (as isotope abundances) of their activities.

Manfred Schidlowski and others have developed detection techniques for measuring carbon, nitrogen, and sulfur isotopes (phosphorous and hydrogen isotopic ratios are not particularly helpful). They have also analyzed the isotope preferences of various organisms. By examining isotopes in appropriate sediments, they have thus been able to approximate when in the fossil record certain types of metabolism (but especially autotrophy) arose (e.g., Schidlowski et al. 1983).

CARBON ISOTOPES

THE TWO most abundant isotopes of carbon are ^{12}C and ^{13}C. ^{12}C is 89.4 times more common on Earth than is ^{13}C. In sediments and in sedimentary rocks where organisms have been fixing the carbon in carbon dioxide autotrophically—that is, making sugars via the Calvin cycle or equivalent—the lighter isotope appears in greater than the usual abundance. This is because RuBP carboxylase, the major enzyme of the carbon fixation step of the Calvin Cycle, favors the lighter isotope (^{12}C). Therefore, sediments and rocks enriched in ^{12}C might reflect autotrophic activity. Important questions are: How well do these carbon isotopes preserve in the fossil record and are they actually reliable indicators of autotrophy by organisms?

When ^{12}C is concentrated in carbonate-rich sediments, the enhanced isotopic ratio seems to remain as the sediments lithify. And because carbonate is often a precipitated by-product of autotrophic activity, the enriched ^{12}C in carbonate rocks probably is a reliable indicator of autotrophy. If organic molecules, however, are buried they gradually degrade to a nondescript organic compound called kerogen. In this case there is a tendency for ^{12}C to be lost because ^{12}C-^{12}C bonds are more easily broken down under thermal stress than are bonds that entail the heavy isotope. This means that carbon isotope ratios in kerogen-rich sediments might not accurately indicate the extent of enrichment that might have occurred, but any enrichment of ^{12}C (above typical background) probably does indicate autotrophy.

The oldest sedimentary rocks with carbon isotope ratios indicative of autotrophy are 3.5 billion years old (from the Warrawoona Group of Australia and the Onverwacht Group of Africa). There may also be evidence as early as 3.8 bya for autotrophy in the Isua rocks of Greenland. In any case, from 3.5 billion years ago to the present there has been a remarkably consistent and uniform record of ^{12}C enrichment in sediments. Therefore, autotrophy appears to be an ancient metabolism, established remarkably soon after the origin of life (approximately 4 bya). This leaves a rather brief window of 300 to 500 million years in which origin of life, followed by fermentative metabolism such as carried out by *Thermatoga* and then autotrophy as in *Chloroflexus'* must have evolved.

While carbon isotope ratios that show enhancement of the light form of carbon are indicators of carbon fixation, they do not differentiate between chemo- and photo-autotrophy, nor between the many different types of photoautotrophy. For this, other evidence must be gathered and analyzed. Nevertheless, as chemoautotrophy requires the presence of some oxygen (to oxidize reduced minerals), it is reasonable to conclude that autotrophy deduced from the most ancient rocks indicates bacterial use of sunlight to provide energy for carbon fixation.

MICROFOSSIL AND TRACE FOSSIL EVIDENCE

THE MICROFOSSIL evidence for autotrophy in ancient sediments, especially those earlier than 2 bya, depends heavily upon isotopic and other data for interpretations. Some samples of archaic rock suggestive of microfossils have traces of organic molecules that may be analyzed. Porphyrin (perhaps derived from cytochromes or chlorophyll) and phytane and pristane chains (perhaps derived from chlorophyll side chains) have been extracted. However, in microfossils of the antiquity in question here, these chemical signs are not considered trustworthy indicators owing to possibilities of contamination from much more recent organisms (Knoll and Awramik 1983).

The simple shapes of ancient microfossils are themselves poor indicators of either identity or metabolism. The oldest microfossils are spherical coccoids found in South African formations and thread-like filaments in Australian formations, both dated at 3.4 to 3.5 bya (Schopf 1976; Dunlop et al. 1978). The isotopic data suggests that these organisms may have been photosynthesizers or at least part of a photosynthetic community.

In modern microbial communities, it is often the photosynthesizers that are more resistant to decay than the rest of the community, because of their tough outer coverings, sheaths, and capsules. It is likely that the parts remaining in many microfossils are also sheaths and capsules, which would indicate that such fossils pertain to photosynthesizers. Which photosynthesizers they might have been is a difficult question.

Chloroflexus, a member of the photosynthetic group of bacteria at the base of Woese's eubacterial tree, grows in long, sheathed filaments. It is therefore as good a candidate as any for some of the filamentous microfossils. Further up on the family tree are many photosynthesizers, both filamentous and coccoid—all possibilities for the identities of ancient microfossils. These include members of the purple bacteria, green sulfurs, and heliobacteria as well as cyanobacteria. All diverged into distinct branches at about the same time. The fact that ancient microfossils are often identified as cyanobacteria may owe to cyanobacteria being the best known and most common bacterial photosynthesizers in modern microbial communities. Early microbial communities might well have been dominated by different taxa of photosynthesizers.

Trace fossils are remains of organismal activity but not the organisms themselves. In macropaleontology, examples of trace fossils are the footprints and feces of dinosaurs. Microbial trace fossils include large, layered structures (stromatolites) formed by the activities of microbial communities and certain oxidized sediments and minerals; both are possible indicators of oxygen-generating photosynthesis.

Stromatolites are layered, domed or bulbous shapes in sedimentary rocks. Living stromatolites are rare, but they can still be found in a few places (like the shallow, saline waters of Shark Bay, Australia). The microorganisms of the stromatolite community, especially photoautotrophs, trap and bind sedimentary particles and effectively cement the particles together forming a stable substrate. In addition to trapping and binding, some members of the community (again, especially the photosynthesizers) induce the precipitation of minerals such as carbonates. Modern stromatolite-building communities are often arranged in layers with cyanobacteria at the top, underlain by a layer of purple photosynthetic bacteria, then by anaerobic heterotrophs (fermenters) (figures 2.6 and 2.7). As sediment is accreted and minerals precipitate, the community moves up to avoid burial, leaving behind a layer of minerals and discarded sheaths (although these sheaths tend not to preserve well in most carbonate stromatolites).

Stromatolite-building communities were extremely successful in the early part of Earth's history. When found in the early fossil record they are generally taken as indicators of photoautotrophy—although some exceptional stromatolite-building communities, based on chemoautotrophy, are found today in thermal springs. Indeed, the growth forms of some fossil stromatolites are suggestive of heliotropy, leaning in the general direction of the sun. Thus they may provide evidence for the orientations of ancient continents. (Kusky and Vanyo 1991; Awramik and Vanyo 1986)

The earliest stromatolites are dated at 3.5 bya in the Australian Warawoona group. That these communities might include photosynthesizers is supported by the isotopic evidence for autotrophy and the presence of what may be microfossils of autotrophs, also at 3.5 bya.

Stromatolites were important structures in the ancient microbial landscape for most of Earth's history, declining in number and size toward the end of the Proterozoic era about 0.6 bya (600 mya) (figure 2.8). Some of the largest stromatolites (truly, microbial skyscrapers) are found near Great Slave Lake in Canada; they are as much as ten meters high and are dated at about 2 bya (figure 2.9). The reason for the decline has been the subject of much discussion among micropaleontologists (e.g., Golubic 1992). Some clues may lie in the nature and distribution of the few stromatolites still existing and actively lithifying today.

For many years stromatolites were known only in the fossil record. But in the late 1950s, Australian paleontologists Philip Playford and Brian Logan stumbled upon a vast community of living

FIGURE 2.6. Microbial mat organisms in captivity (growing in a Winogradsky column) demonstrate the distinctive layering patterns found in the field as well as in the fossil record. Winogradsky columns are a method for culturing bacterial communities in the lab. For example, by supplying water, a source of cellulose (paper towel) and a little $MgSO_4$ (epsom salts) as a source of sulfate, a marine sedimentary community will thrive and even form a distinctive layered pattern in a container. Photography by W. Krumbein.

stromatolites while exploring a hypersaline lagoon in Shark Bay, Australia As Golubic (1992) describes this historic event, the two scientists "thought they were dreaming" when they encountered "a landscape of living stromatolites." Some of these layered mounds are up to 60 cm high and continue to accrete and lithify several millimeters per year. Cyanobacteria are the principal photosynthetic agents and accretors. Since the Shark Bay discovery, other modern stromatolitic landscapes have been identified. These include Solar Lake in the Sinai Desert and the salinas (salt flats) on the west coast of Baja California, Mexico.

Why are stromatolites, which were apparently so successful for

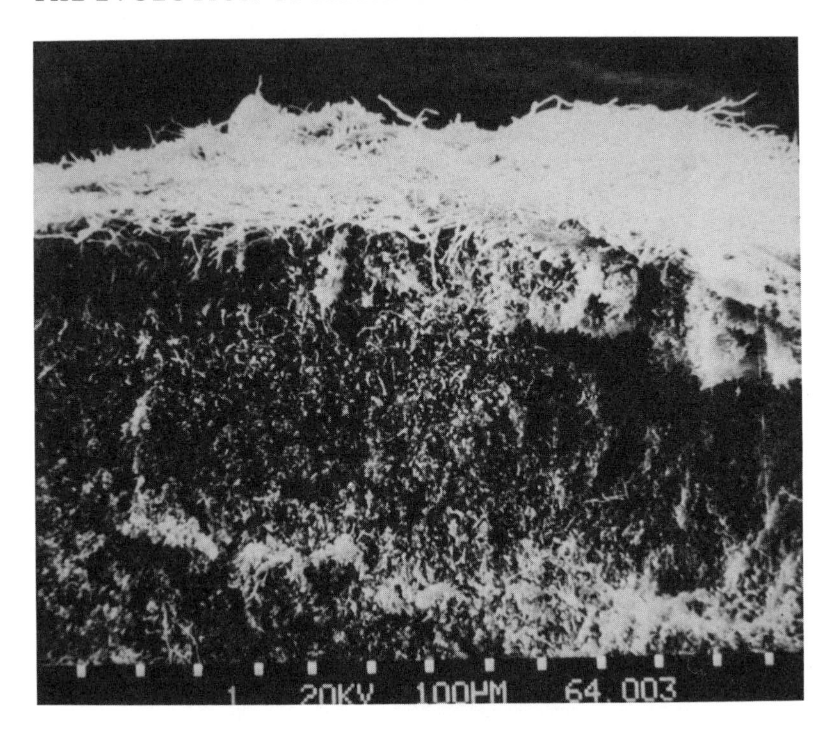

FIGURE 2.7. A scanning electron micrograph of a cyanobacterial community shows the matlike structure composed of tightly woven filaments and bound sedimentary particles. Photography by W. Krumbein.

three billion years of Earth's history, so rare today? The decline ncoincides with the origin and increasing diversity of invertebrate animals, notably grazing and burrowing herbivores. These animals may have become so successful at profiting from the enormous photosynthetic capacity of microbial communities that no extensive accretions and lithifications could occur. In fact, some grazing invertebrates such as molluscs are equipped with rasplike structures (radula) for grinding through the calcium carbonate precipitates of photosynthesizers. Environments conducive to the growth of stromatolites after the Cambrian "explosion" of marine animals were thus those especially inhospitable to invertebrates, such as hypersaline or thermal pools. However, the evolution of predators is not considered to be the entire explanation for the decline of stromatolites. It may also be that certain environmental conditions, such as super-saturated solutions of minerals (that might be expected in hypersaline pools) are especially conducive to lithification.

FIGURE 2.8. Stromatolites from Ballston Spa, New York, approximately 0.5 billion years old, the waning years of the great age of stromatolites that had lasted for about 3 billion years (3.5 to 0.5 bya). Photograph by A. Breitbart and R. Cooper.

The largest extant stromatolites are branching towers, some an impressive forty meters high found in an alkaline lake in Turkey (Kempe et al. 1991). This lake is saturated with carbonate minerals, thereby promoting accretion and lithification. Perhaps some change in the conditions of dissolved minerals at the end of the Proterozoic promoted the demise of stromatolites. Both calcium carbonate and silica (the major lithifying minerals in stromatolites) are scavenged and used by shell-forming organisms of all types (e.g., diatoms use silica, most molluscs use calcium carbonates). Perhaps an increase in shell-making organisms, which is certainly reflected in their sud-

FIGURE 2.9. Skyscraper stromatolites two billion years old from Great Slave Lake, Canada. Courtesy of the Geological Survey of Canada.

den appearance in the fossil record at about 0.6 bya (600 mya), was the crucial evolutionary event. For example, gigantic coral reefs were sites of massive calcium carbonate precipitation. Perhaps those upstart organisms were simply better at capturing available carbonates than were the older stromatolitic organisms.

Finally, the evolution of highly efficient forms of heterotrophy, especially respiration, may have enhanced the "consumption" of stromatolitic structures from beneath (or inside). The deep layers of living stromatolites include heterotrophs that feed on the exudates and dead bodies of the autotrophs left behind as accretion continues. Usually these heterotrophs are anaerobic and their consumption is relatively inefficient and slow. Golubic (1992) noted that introducing oxygen into the deep layers, through cracks for example, greatly enhances the rate of consumption by oxygen-using respiring heterotrophs. Thus the evolution of respiration may also have enhanced the abilities of heterotrophs in general (not just grazing invertebrates) to consume the photosynthesizers before extensive lithification could occur.

OXYGEN IN THE FOSSIL RECORD

EVIDENCE OF molecular oxygen in the atmosphere along with oxidized minerals in the sediments are indicators of two important types of metabolism: oxygen-generating photosynthesis and oxygen-consuming respiration. Most of the free oxygen in Earth's atmosphere is the waste product of oxygenic photosynthesis, performed by cyanobacteria and various photosynthetic eukaryotes such as algae and plants; only a small fraction of the free oxygen is produced in the upper atmosphere from the photolysis of water (Walker et al. 1983). When oxygenic photosynthesis first evolved, oxygen would have reacted promptly with whatever reduced compounds were in the vicinity. On early Earth the soils, rocks, and waters were mostly reducing. Therefore, it is widely believed that oxygen did not begin to accumulate in the atmosphere until all available, exposed substrates had been oxidized. The time period during which various minerals were oxidized can be roughly established by looking for certain patterns of oxidation in the fossil record.

Banded iron formations are one such pattern. They are layered sedimentary rocks with red, oxidized-iron strata alternating with gray, iron-poor strata. Oxygenic photosynthesis, occurring either on a diurnal or a seasonal cycle, might have produced the oxidized iron layers which would have occurred as a rustlike precipitate, in a shallow body of mostly anoxic water. The oldest banded iron formation is 3.8 bya but most are found between 2.2 and 1.8 bya (Walker et al. 1983). It is not known whether all banded iron formations were formed by the same process, and thus Walker and his colleagues are reluctant to include the odd, oldest one at 3.8 and another at 2.5 bya with the more recent cluster. A safe interpretation might be that significant amounts of oxygen were produced (on a cyclic basis) by photosynthesizers as early as 2.2 bya.

At about 1.7 bya, red beds (sedimentary rocks uniformly enriched with oxidized iron) became common, suggesting that at this point there were no more anoxic surface environments left to be oxidized. Approximately 1.7 bya may therefore be a date at which oxygen began to accumulate in the atmosphere, although some red beds date as early as 2.1 bya. Is 2.2 bya the earliest date for oxygenic photosynthesis? Probably not, given other lines of evidence that would push back the date to 2.8 bya, if not earlier. Hayes (1983) notes that changes in the carbon cycle, deduced mostly from isotope data, suggest a presence of oxygen presumably of photosynthetic origin. For example, later than 2.8 bya kerogens (the highly modified remains of buried organisms) become depleted in ^{13}C. This is an indirect indication for a cycle involving methane-producing bacteria and methane-oxidizing bacteria, both of which require some measure of free oxygen.

NITROGEN ISOTOPES

THE LIGHTER isotope of nitrogen, ^{14}N, is 277 times more abundant than ^{15}N. Prokaryotes that either nitrify or denitrify prefer the lighter isotope and therefore increase its abundance in sediments in which they are active. The nitrification reaction is performed by nitrifying bacteria that are aerobic chemoautotrophs.

These organisms fix carbon dioxide using the energy from ammonia or nitrite. Some are also facultative heterotrophs. The general nitrifying reaction used to get energy for carbon fixation is:

$$\underset{\text{(ammonia)}}{NH_4^+} + 1\tfrac{1}{2}\,O_2 \rightarrow \underset{\text{(nitrite)}}{NO_2^-} + 2H^+ + H_2O$$

$$\underset{\text{(nitrite)}}{NO_2^-} + 1/2\ O_2 \rightarrow \underset{\text{(nitrate)}}{NO_3^-}$$

The ATPs formed are used as an energy source to make NADPH and to fix carbon dioxide. For a description of how the carbon fixation (Calvin) cycle works and how the ATPs and NADPHs are used, see chapter 7.

The denitrifying reaction is performed by a diversity of heterotrophic prokaryotes including *Pseudomonas* and *Bacillus*. A simplified reaction using glucose as the food source and nitrate as the oxidizer is:

$$C_6H_{12}O_6 + 4\ NO_3^- \rightarrow 6\ CO_2 + 6\ H_2O + 2\ N_2$$

(glucose) (nitrate) (gaseous dinitrogen)

Denitrification bacteria use nitrate in this reaction as a terminal electron acceptor, analogous to the way that oxygen is used in respiration, only this occurs anaerobically. While denitrification enzymes favor the lighter isotope of nitrogen, the gaseous waste product (molecular nitrogen) is unlikely to remain in the sediments.

Some bacteria also need to fix atmospheric nitrogen to ammonia if more suitable forms are unavailable. The atoms in atmospheric nitrogen, N_2, are bound together in a near-unbreakable triple bond. Nitrates, nitrites, and ammonia are therefore far preferable sources for obtaining nitrogen essential for building DNA and proteins. Nitrogen fixation is a metabolic activity accomplished by various bacteria, such as *Azobacter and Anabaena,* and it somewhat favors lighter nitrogen. The general equation is:

$$N_2 + 3\ H_2O \rightarrow 2\ NH_3 + 1\tfrac{1}{2}\ O_2$$

(gaseous dinitrogen) (ammonia)

Other forms of nitrogen metabolism do not seem to leave fossil signatures. For example, when organic nitrogen compounds such as urea and ammonia are excreted as waste products they generally show little evidence of fractionation.

When nitrification, denitrification, and nitrogen fixation are combined they form a nitrogen cycle. Although abiotic forces (such as lightning-induced nitrogen-fixation) contributed to the chemical transformations of nitrogen before life got involved, the biotic components of the nitrogen cycle served to greatly enhance the efficiency and flow of the overall cycle. These key biotic components of the nitrogen cycle probably evolved as early as 2.5 billion years ago because kerogen occurring in rocks of that age is noticeably enriched with the lighter nitrogen isotope.

NITROGEN ISOTOPES AS INDICATORS OF OXYGEN

OXYGEN IS a necessary part of the nitrogen cycle which appears to have been in place at 2.5 bya. Thus oxygen seems to have been present in sediments and waters by about 2.5 bya. Indeed Knoll (1992) speculates that 2.8 bya is the lower limit for the evolution of eukaryotes because it was then that sufficient atmospheric oxygen (about 0.5 percent) would have been available for the origin of respiration. Here, however, arises a sticky problem. Woese's eubacterial tree (figure 2.1) shows cyanobacteria (the oxygenic photosynthesizers) and various respiring heterotrophic groups branching off almost at the same time. If the evolution of oxygenic photosynthesis was followed quickly by the evolution of respiration, respiration could have served as one of the earliest "removal systems" for the then widely toxic oxygen, leaving a less accurate fossil record in those oxidized sediments. That respiration might have evolved quickly is not such a strange idea, in that all of the crucial semes would have been in place. Then too, the act of evolving respiration seems to have been comparatively easy in that there is evidence that it occurred several times in different branches. Furthermore, the data that enable the rise of oxygen to be dated do not take into account the possibility of localized microenvironments of oxygen, which may have occurred long before detectable oxygen accumulated in the atmosphere.

It may be safest to say that oxygenic photosynthesis and perhaps even respiration evolved at least by 2.8 bya and perhaps even earlier. How much earlier may be impossible to ascertain. The photosynthetic-looking microfossils and the well-established stromatolites dating from 3.5 bya may indeed be evidence for an earlier nonoxygenic photosynthesis.

SULFUR ISOTOPES

THE SULFUR isotopes provide circumstantial evidence for oxygen in that a complete sulfur cycle implied by the data would require the presence of oxygen. The lighter sulfur isotope, ^{32}S, is 22 times more abundant than the heavier isotope ^{34}S. One particular type of bacterial metabolism significantly favors the lighter isotope of sulfur and is likely to leave a detectable organic product in sediment. This metabolism is sulfate reduction by heterotrophic desulfovibrio bacteria, which use sulfate as a terminal electron receptor analogous to the way in which oxygen is used in respiration. This kind of sulfate reduction occurs in anaerobic sediments where the hydrogen sulfide waste product may react immediately with ubiquitous iron to form iron sulfide minerals such as pyrite. The first appearance of pyrite enriched in light sulfur appears in rocks 2.7 billion years old—the banded iron formations and sulfide deposits of Michipicaten and Woman River in Canada. Therefore sulfate reduction may have appeared around this time or earlier.

There is minor enrichment of light sulfur isotopes when sulfate is incorporated by organisms into proteins and when hydrogen sulfide is used by photosynthetic sulfur bacteria, or when sulfide minerals are used by chemoautotrophic bacteria. These metabolisms, however, do not seem to have left an identifiable mark in the fossil record.

SUMMARY OF THE FOSSIL EVIDENCE OF METABOLIC EVOLUTION

TO SUMMARIZE, the apparently rapid evolution by 3.5 bya of photosynthesis, perhaps of a *Chloroflexus*-type bacterium (although this cannot be accurately determined), was followed by oxygenic photosynthesis and respiration such that all may have been accomplished by 2.8 bya. This implies an astonishingly compressed interval for such major evolutionary events. And a compressed time line is decidedly evident in the family tree based on rRNA data. That is, the evolution of fermentation (which has no known fossil signature), followed rapidly by photosynthesis, seems to have been enough in the way of ground work for all the branches of eubacteria to diverge almost at once, with variations on heterotrophy and autotrophy. A reasonable conclusion is that by about 2.8 bya all of the eubacterial types were present except perhaps some odd branches that are exclusively parasites of eukaryotes.

Two and a half billion years ago is also the earliest end of the range estimated by Knoll (1992) and others (chapter 1) for the origin of eukaryotes. Thus the stage is set for the events that are the focus of this book: the assembly and further evolution of eukaryotic cells. But first, because prokaryotes have another important trait drawn upon by eukaryotes, we shall devote one more chapter to this essential background.

III

Horizontal Gene Transfer: Cementing Relationships

ONE OF the most intimate ways that two organisms may associate in a symbiotic relationship is by sharing genes. Horizontal transfer of genes, an exchange of genes between partners, has had a long evolution in bacteria. When some bacteria associated in relationships that would establish eukaryotic cells, horizontal transfer was probably a crucial mechanism for making the symbioses obligate. Sharing genes can make relationships more streamlined and efficient. It also cements those relationships by preventing the partners from easily separating, especially when important genes have been moved. This chapter describes the mechanisms of horizontal transfer and how those mechanisms have bound together the symbiotic parts of eukaryotic cells.

The genomes of organisms are enclosed within the protective membranes of the cell and (in eukaryotes) also within the protective membranes of the nucleus and organelles. Genomes, as such, have the appearance of protected entities that maintain their integrity, save for genetically prescribed crossings inherent in sex. This isolation of the genome is a premise upon which most of the early work in genetics was based, and it has long been a part of the canon of biology textbooks. The traditional view regards genetic mobility only within the bounds of reproduction: the duplication and passage of genes during cell division. Genes, of course, could also be packed into gametes (sperm and eggs) and sent out to combine sexually with other gametic genes. However, the canon did not allow for the possibility that genes might travel in whole or in pieces from cell to cell, both intra- and interspecifically, and that the genome of an organism might be fluid.

Some of the earliest work on the topic of moving genes was that of Barbara McClintock and her colleagues, who discovered in the 1940s and 50s a particular type of moving gene, "transposons" or "jumping genes," in corn plants. These jumping genes had the ability to move into and out of a chromosome and to insert themselves in various places along the same or another chromosome, including the middle of another gene. These rather intrusive genes manifested themselves by altering some of the traditional Mendelian traits that one expects to find in simple crosses of corn.[1]

This fascinating work was virtually ignored for two or three decades as geneticists held to the belief that genes were mostly stable and stationary entities, faithful within a species. Such a presupposition is a natural outgrowth of the fact that in human medical genetics the detection of movements of genes is nearly always correlated with a genetic defect. The surprise was that jumping genes turn out to be not so exceptional or pathological and that there are many mechanisms by which genes and parts of genes move from chromosome to chromosome, cell to cell, and species to species. These movements may be major mechanisms in evolution, especially the evolution of prokaryotic cells. At this point even specialists can only guess at how extensive the gene movement or "horizontal transfer" might be. The problem is that the movements of genes can be traced only if they carry or leave behind some signature of the event. Genes

[1] For example, if corn with colored kernels (i.e. with the genotype CC) is crossed with corn of colorless kernels (cc), we expect only colored, Cc, offspring in the first generation, because C is dominant over c. However, if a jumping gene is moving about in the genome, then some of the offspring kernels may look colorless (representing C genes that have been invaded and rendered nonfunctional by a jumping gene), or they may be speckled, with color representing the occasional departure of a jumping gene.

that make a "clean" transfer, leaving no detectable clue may be (with present technology) untraceable.[2]

The goals of this chapter are to describe the major mechanisms by which genes are known to move between organisms and recombine and to give specific examples of genes that, thanks to distinctive clues, are known to have made a jump. The case will be made that the horizontal transfer of genes has been a primary mechanism in the evolution of prokaryotic cells and later in the evolution of complex cells (as "communities" of prokaryotic cells).

BACTERIAL GENE MIXING AND REPAIR

BACTERIA HAVE their own particular version of sex or gene mixing, a mechanism that is so fundamental to the properties of DNA that it has probably been going on since the early evolution of the prokaryotes. This sexual activity is simply genetic recombination.

DNA strands are in general stable molecules, and this is one of their most useful properties as informational molecules in cells. But they are also susceptible to mutations, changes and breaks in the nucleotide sequence, caused by mistakes in DNA replication or by outside agents such as ultraviolet light. In fact, it may have been ultraviolet components of sunlight penetrating the atmosphere of early Earth that provided the strong selection pressure for systems that repair mutations of DNA.

Earth's early atmosphere had little or no free oxygen and therefore no ozone layer to protect organisms from ultraviolet radiation. Not until about two billion years ago did sufficient oxygen (and ozone) accumulate in the atmosphere, as a waste product of oxygen photosynthesizers, to diminish the threat of ultraviolet light. Before the rise of the ozone shield, organisms would have evolved mechanisms to protect their DNA from ultraviolet light (Margulis and Sagan 1986). One of these mechanisms probably came rather naturally to DNA: recombination. For another property of DNA, in addition to mutability, is its ease of repair, owing to its molecular existence as pairs of base sequences entwined in a double helix. The undamaged,

[2] The jumping gene may, however, provide a clue if its sequence of bases looks markedly different from other sequences in the same taxon and similar to those of a distant taxon (Doolittle et al. 1990). But the fact that genes may move more than once is an additional complication.

adjacent strand can serve as template for reconstructing the damaged sequence. Thus repairs can be easily made by replication enzymes, and in fact may look like localized replication events.

In one type of repair called recombinational repair, adjacent double strands of DNA may be used to patch a mutant DNA strand. This type of repair occurs just after DNA replication and it targets a specific type of mutation caused by ultraviolet light, the thymine dimer. Thymine dimers form bonds with each other on the same strand rather than base pairing with adenines on a complementary strand. Thus a strand that should look like this:

```
ATTCG
|||||
TAAGC
```

may, after a mutation, look like this:

```
AT—TCG
|    ||
TA  AGC
```

After DNA replication a hole is left in the new strand where DNA polymerase was unable to make a complement for the thymine dimer.

```
AT—TCG
|    ||
T    GC
```

An adjacent double-stranded DNA molecule, presumably identical in sequence (except for the mutation) as it is the other replication product, supplies the correction by recombining with the mutant DNA and patching the correct sequence into the hole. This leaves the normal DNA with a hole, but one that is easily patched to complement the undamaged sequence on the other strand.

This recombinational repair was almost certainly a precursor for bacterial sexuality. When used strictly as a repair mechanism immediately after DNA replication so that the appropriate complementary strand is available, the event is not particularly sexual. However, when the adjacent strand is not an identical one (for DNA will casually pair with imperfect complements) and especially if the adjacent strand is from a different genome, then this is sex, by the most basic definition. Sex is the recombination of DNA from two or more sources, a genetic mixing of genomes. All that is needed to convert recombinational repair to sex are mechanisms for placing genomes of different individuals within easy access of each other.

Even today, sex in bacteria, and in most organisms for that matter, is not necessarily linked to reproduction. Reproduction and growth involve the replication of entire genomes, and usually cell division. Reproduction is fundamental to life and, along with self maintenance (autopoiesis), defines living systems. A reproducing organism produces two or more identical or nearly identical offspring. In contrast, sex as the mixing of genomes is not essential to life and is by no means a requirement for reproducing organisms—unless it has become linked with reproduction, as is the case with some types of animal sex.[3] In fact, researchers have searched long and hard for evidence that sexual organisms have any particular adaptive advantage over asexual ones, and have found almost nothing (Margulis and Sagan 1986).

THE ADVANTAGES OF SEXUALITY

ONE EXCEPTION in which sexuality does confer a fitness advantage is in environments with strong selection pressures for organisms with certain combinations of genes that happen to be dispersed among two or more different organisms. It may be that such environments exert this selection pressure for the specific mechanisms by which bacterial genomes come together. For example, on a petri plate containing the antibiotics penicillin and tetracycline, a mixture of a population of tetracycline-resistant bacteria and a population of penicillin-resistant bacteria would be expected to leave no offspring. However, if these two populations have mechanisms for sex, a few recombinant progeny that contain the genes for both tetracycline and penicillin resistance will be strongly selected and will leave many descendants.[4]

While reproduction yields two or more offspring, sex often results in exactly the same number of individuals, but with new gene combinations. Extant bacterial groups abound with mechanisms for getting genes together. Inventions to facilitate sex include structures

[3] Sex can also yield fewer individuals, as in the case of two *Chlamydomonas* green algae that fuse sexually, yielding one individual; the mitotic steps that subsequently occur are what yield two or more offspring.

[4] In an environment free of antibiotics, sexuality may still be occurring but with no particular advantage for doubly resistant recombinants, and perhaps even some disadvantages owing to the burden of carrying excess and unneeded DNA.

like fimbriae by which cells adhere to each other and pili (tubelike structures) through which DNA can pass. Successful recombinants will survive but so too will organisms that are motile and can move away from the stress or that form protective cysts and can wait out the disaster. Sexuality therefore is one of many strategies for bacteria that find themselves under environmental stress.

VIRUSES AS SEXUAL AGENTS

VIRUSES CAN act as sexual agents for bacteria and other organisms. These genetic entities consist of a DNA or RNA genome, sometimes enclosed within a protein coat, which depend entirely upon a host cell for their replication and propagation. They are, therefore, not "alive" by the basic definition of life, in that they do not reproduce or perform autopoietic functions unless they have the use of a full set of cell autopoietic (self maintenance) and reproductive mechanisms. Viruses are intrinsically sexual in that they have the means to get inside of cells and expose their DNA or RNA to the cellular enzymes, which will then maintain and replicate it. In doing so they also expose their DNAs to those of other viruses that might be invading the same cell or to the DNA and RNA of the host.

Recombinations can thus occur between the viral DNAs and between viral and bacterial DNAs and, if the virus carries bacterial DNA from some previous encounter, between bacterial DNAs. The viruses of bacteria may have been evolving almost as long as their bacterial hosts (e.g., four billion years) and therefore are remarkably fine-tuned to their hosts and often extremely efficient in the few functions that they require to survive. In a sense, the entire function of a virus, from its own point of view, is to become integrated genetically with a host so that the viral DNA is treated as host DNA or treated even better. (Many pathogenic viruses, in fact, cause their hosts to destroy their own DNA and to devote resources to the well-being of the viral invader.) The viral strategy often includes recombining the viral DNA with the host DNA.

These ultimately sexual entities, the viruses, have had four billion years of practice in mixing DNA, and in a sense are using a very advanced form of sex. This advanced, highly specialized sexuality also places many viral pathogens at the frontiers of medical science;[5] the challenging medical question is what to do with pathogens that become, in a sense, part of their hosts, but that also readily change their genomes (and identities) to evade the defense systems of the host.

SEXUAL PROMISCUITY (HORIZONTAL GENE TRANSFER)

THUS FAR we have deliberately omitted mention of the specificity of sexual activity between bacteria, with or without viruses. Traditional definitions of sex, especially those centered on mammalian sex, have always focused on the species-specific nature of the activity. Mammalian species are, in fact, defined on the basis of their abilities to interbreed exclusively within their own unit. In expanding the basic definition of sex to include all organisms, we and others (e.g., Sonea and Panisset 1983; Margulis and Sagan 1986) do not regard species specificity as a diagnostic feature of sex. Rather, we include under the heading of sex any mixing of genomes. We also have not specified how extensive the mixing must be, for we include recombinations of parts of genomes, and the nonreciprocal mixing which often happens in bacteria, as well as the integration of entire genomes.

The expanded definition of sex as the nonspecific mixing of all or parts of genomes gives a broad base from which all of the numerous types of sexuality evolved. Beginning with the first eukaryotes, a group called the protoctists, several new types of sexuality evolved. Some of these variations became characteristic of the later eukaryotic kingdoms: plants, fungi, and animals. While sex has, in fact, become rather limited (that is, species specific) in many eukaryotes, it seems to have become even more promiscuous in the prokaryotes. The mixing of genomes across species boundaries (sometimes called horizontal gene transfer) occurs between bacteria, as well as between bacteria and the four eukaryotic kingdoms.

[5] Not all viruses are pathogens; in fact very few may be. However, pathogenic viruses are by far the best studied because our paradigm for viral study is to identify damage to host cells (pathogenicity) rather than to look for viruses themselves.

MECHANISMS OF HORIZONTAL GENE TRANSFER

THE THREE basic steps for successful gene transfer between genomes begin with a gene duplication event. This is followed by physical transfer of one copy of the gene from one genome to another. Sometimes the replication event and the transfer event are combined in one step, such as when a virus picks up and duplicates a gene. Transfers that occur to unduplicated genes are most likely to be lethal to the cell that loses its sole copy of the gene. After a translocation event, the recipient cell must somehow activate the gene, in order for the gene to become functional. To complete the transfer, thus, the gene must become coordinated with the various regulatory mechanisms that the recipient cell uses to transcribe and translate DNA. The accomplishment of all three of these steps is unusual, and thus successful gene transfer is a rare occurrence. Nevertheless gene transfer has occurred many times in the evolution of organisms and must be considered an important cause of genetic variability. (Further details on this sequence of steps and the mechanisms of horizontal gene transfer are in appendix C.)

Although realization of the widespread nature of interspecies gene transfer is recent, mechanisms that can power this phenomenon have been known for some time. There are three major mechanisms by which genes and parts of genes may move from one cell to another. These are *conjugation,* by which a donor bacterium transfers some or all of its genes to a recipient, *transduction,* by which the gene transfer is accomplished by a viral intermediate, and *transformation,* in which cells pick up pieces of DNA from their environment. In all three mechanisms, it is necessary for the pieces of foreign DNA to be retained by integration into the genome of the recipient host or in a piece of satellite DNA (a plasmid) in order to be passed down to succeeding generations. In fact, the major mechanisms of natural gene transfer are also the mechanism of artificial gene transfer, as these mechanisms and many variations of them have been standardized and turned into everyday laboratory techniques forming the foundations for recombinant DNA technology (genetic engineering).

A key question, from an evolutionist's point of view, is to what extent such processes for gene exchange (horizontal gene transfer) are present in the natural environment. Evidence from several lines of study suggests that horizontal transfer has been and continues to be an important mechanism in bacterial evolution. For example, Kraut et al. (1989) examined one of the genes involved with carbon monoxide oxidation—a type of autotrophy usual in organisms that can also live heterotrophically. The gene was found on the main chromosome in *Arthrobacter* and *Bacillus,* on a plasmid in *Alcaligenes* and *Azomonas,* and at either of these locations in two species of *Pseudomonas,* depending on the particular strain of a given isolate. Clearly, this is a dynamic gene that has moved several times, and which may well have spread through these bacterial genera by some mechanism of horizontal transfer. The fact that carbon monoxide oxidation is usually an alternative form of metabolism performed by a heterotroph, rather than an obligate characteristic, means that the genes may be especially susceptible to the risks (and occasional losses) of transfer. That is, accidental loss of a gene for carbon monoxide oxidation would not be lethal for its owner, if alternative forms of metabolism were available.

Plos et al. (1989) examined 229 strains of *Escherischia coli* and found that even some isolates that supposedly represented the same clone had differences in the positions of "pap" genes (which produce part of a structure used in conjugation called a pilus). This strongly suggests that the pap genes get handed around rapidly in *E. coli,* a phenomenon that may be common in other genes as well.

A group of Russian researchers, Mindlin et al. 1990, recently performed an experiment that would be conceptually challenging for many western microbiologists trained in the art of monoculture and sterile techniques, but which is quite natural in scientific descendants of A. P. Winogradsky (the Russian inventor of the "Winogradsky Column" for culturing whole communities of bacteria). *Acinetobacter, Escherischia,* and *Pseudomonas* were cultured together and observed for their tendency to pass around a particular plasmid DNA across the species—indeed, genus—boundaries. They did this quite readily and in frequencies similar to or better than the natural background mutation rates of one in 10^5 to 10^6cellular divisions. The mechanism of the transfer was not determined.

Coughter and Stewart (1985) reviewed evidence that confirms that the three major mechanisms of transfer in bacteria—conjugation, transduction, and transformation—all occur in natural settings such as soils.

Sonea and Panisset (1983) claim that continued access to horizontal gene transfer may be a principal difference between prokaryotes and eukaryotes, and that the "genetic unity" experienced by prokaryotes leads to cooperative associations (symbioses) or intricate community structures. They consider separate bacteria taxa to be related and functional parts of a single and worldwide gene pool "united by the possibility of sharing genes." Sonea and Panisset go so far as to say that bacteria might tend toward a "genetic incompleteness" (lack of metabolic flexibility) necessitating cooperative associations. This interdependency might be a state that is both caused by horizontal transfer and perpetuated by it; that is, organisms that can share genes easily may be less likely to maintain full complements of genes themselves.

The most extensive cases of horizontal gene transfer known are those that occurred during the evolution of the eukaryotic cell. As much as 98 percent of the genes of plastids and mitochondria have been transferred into the nuclear DNA, or lost, and the bacterial tendency toward sharing of genes could have been a strong preadaptation for this trend.

OUT OF THE LOOP

WHEREAS HORIZONTAL gene transfer may be a frequent phenomenon in prokaryotes, there are some bacterial groups that may be outside of the gene-transfer loop (Syvanen 1987). These include those extremely rare prokaryotes that have modified parts of the "universal" genetic code (the code by which information in DNA sequences is translated to make protein sequences). Mycoplasms, which are wall-less parasites related to gram-positive bacteria such as *Bacillus,* use a modification of the "universal" stop codon to code for an amino acid, tryptophan. This would ordinarily be a lethal mutation, but it was somehow fixed in mycoplasms (Yamao et al. 1985). A slight difference in language may be enough to isolate mycoplasms from horizontal gene exchange. Mycoplasms do not, in fact, pick up plasmids from other bacteria, not even useful ones such as plasmids conferring resistance to antibiotics (Syvanen 1987). Furthermore, mycoplasms have an unusual degree of variation in their rRNA sequence when compared to other bacteria and when compared within their group, mycoplasm to mycoplasm. While this could indicate an elevated mutation rate in mycoplasms (Woese 1985), it might also be attributable to their being out of the gene transfer loops that tend to homogenize certain sequences (Syvanen 1987).

The most important example of variation in the universal code has occurred several times in different lines of mitochondria. The change apparently followed a considerable number of gene transfers between the mitochondrion and the nucleus, and has had the final effect of isolating mitochondria genetically and thus preventing complete transfer of the mitochondrial genome into the nuclear DNA. (This change will be discussed more fully in chapter 6.)

CEMENTING RELATIONSHIPS BY HORIZONTAL GENE TRANSFER

WITHOUT DOUBT, evolutionary innovations for horizontal transfer of genes were themselves major events in early bacterial evolution, along with the evolution of metabolism. But these transfer methods have, moreover, continued to serve as evolutionary mechanisms in their own right. Horizontal gene transfer may be the "glue" that holds together bacterial associations, and thus may have been a key preadaptation in the symbiotic associations that led to complex cells.[6]

There are many examples of gene exchange between prokaryotes and eukaryotes that go on today. The examples reviewed here are those reminiscent of the sort of gene transfer thought to have occurred extensively between hosts and symbionts during eukaryotic evolution. It is thus a preview of events that will take center stage in chapters 4 through 8.

[6]There is a simulation game called "Prisoner's dilemma" that demonstrates how the behavior of a self-interested individual might lead to overall cooperative behavior between individuals. In one version of this game the play involves the periodic exchange of money for goods in a secretive manner. Thus the participants never meet and they make their exchanges in closed bags, not to be opened until the exchange is complete. Fear of receiving an empty bag and a desire to give an empty bag (thus making a profit) are two of many motivating factors in the game. A tournament, however, yielded surprising results: the most efficient and beneficial way to play is *fairly,* with little or no cheating. There are implications for many fields including biology and for the establishment of intricate symbionts. It suggests that cooperative interactions may be selected in a system that favors efficiency and maximum yield of benefits. For more on this intriguing idea, see Axelrod (1984) and Hofstadter (1985).

The gene for isopenicillin N synthetase, which enables organisms to produce penicillin, appears to have been transferred from the prokaryote *Streptomyces* to eukaryotes, the fungi *Aspergillus* and *Penicillium* several times (Weigel et al. 1988; Landan et al. 1990; Penalva et al. 1990). Eukaryotes may also exchange genes among themselves, although they usually require a more elaborate transfer mechanism than do transfers originating with prokaryotic donors. For example, different species of fruit flies (*Drosophila*) exchange genes via a tiny mite that lives parasitically on their bodies (Houck et al. 1991). This explains in part why gene sequences are not always reliable in working out family trees for fruit flies and other eukaryotes and why some sequences that don't fit with the rest of a phylogeny must be thrown out of the analysis. *Drosophila melanogaster* and *D. nebulose* are closely related species, and yet some of their genes (having arrived by horizontal transfer) are from widely different sources (Hagemann et al. 1990).

Vivid as these eukaryotic examples are, most horizontal transfer occurs between prokaryotes and is fairly limited between eukaryotes (Stachel and Zambryski 1989; Heinemann and Sprague 1989). But *within* each eukaryotic cell, gene flow has been considerable between the mitochondria, plastid, and nuclear genomes (all formerly prokaryotic). In fact, the genetic analysis of a eukaryotic cell can be viewed as the population genetics of that cell, and properly should include the various components of population genetics: the frequency of alleles (gene types) in the population, the mating structure, selection pressures, etc. (Birky 1978). On the level of the entire (often multicellular) eukaryote there have evolved barriers to easy horizontal transfer, including physical ones such as the protective layers covering many organisms. Eukaryotes resemble genetic islands mostly isolated (when compared to prokaryotes) from overly frequent and haphazard horizontal transfer.

In any intimate association between species, especially prokaryotic species, horizontal transfer may be the glue that bonds relationships. One of the best examples of this is the gene exchange that must have occurred in the symbiosis between leguminous plants and rhizobium bacteria. The plants culture the bacteria in special oxygen-free nodules in their roots; there, the bacteria fix nitrogen, some of which is drawn upon by the plant for constructing amino acids. Nitrogen fixation requires an oxygen-free environment, which is maintained within the nodule with a type of hemoglobin and cytochromes that bind up free oxygen. In the case of the symbiosis between soybeans and *Bradyrhizobium*, the bacteria have the gene to make one of the precursor molecules for the heme of hemoglobin; the plants have the other (Sangwan and O'Brian 1991). A gene transfer is believed to have occurred in the history of this symbiosis, resulting in the somewhat awkward shared pathway. The benefits of sharing must outweigh the awkwardness in that the organisms so engaged have a strong mechanism for staying together as well as for controlling the growth of their partners and keeping in synchrony.

Mitochondria and plastids are also excellent examples of symbionts that share coding with the nucleus for certain essential structures, thus cementing the relationships that will be described in chapters 6 and 7. A complete or nearly complete transfer of genes might explain the difficulties (described in chapter 8) of finding evidence to support hypotheses of symbiotic origins for the motility organelles as well as some other organelles, such as hydrogenosomes and peroxisomes.

IV

The Eukaryotic Host Cell

IN THIS chapter the nature of the original host cell that acquired mitochondria and plastid symbionts and perhaps motility symbionts will be described. More details on the acquired symbionts themselves and the sequence of events and selection pressures leading to the assembly of the first symbiotic, complex cell will be found in later chapters on mitochondria, plastids, and motility organelles.

ARE THE ARCHAEBACTERIA A GOOD COMPARISON GROUP?

IDENTIFICATION OF an appropriate modern prokaryote as the close relative of the original eukaryotic host cell is a challenge. Woese's evolutionary tree based on rRNA sequence data (chapter 1) shows that the Archaebacteria are closer than the Eubacteria to the line of prokaryotes that became the eukaryotic host cells (Eucarya). Lake's interpretation of some of the same sequences and others places some of those archaebacteria into a group even closer to the eukaryotes, the eocytes. However, it is important to note that all three or four of these ancient prokaryotic lines diverged close to four billion years ago and have been evolving along their own lines (except for symbioses and horizontal gene transfers) ever since.

Any Archaebacteria or Eocyte lineage, no matter how close to the ancestral branching point at which the Eucarya split off, is going to be a challenging mix of ancient traits. Some traits will be common to both branches and some unique to Archaebacteria—most of which evolved in very unusual environments, such as hot springs and salt flats. In spite of the difficulties (pointed out by many, including Doolittle 1987), it is the Archaebacteria that we will be using as one of the only sources of information about the original eukaryotic host cell.

There are apparently no Eucarya left except for those that made the transition to nucleated (and mostly symbiont-filled) eukaryotes. Those Eucarya that made the transition may have outcompeted any that did not, and there is an additional possibility that relatively few remained free of symbionts. Some characteristics of the original host line might have been highly conducive to the uptake of internal symbionts. Thus a good part of the discussion in this chapter will focus on Archaebacteria, represented by such extant genera as *Thermoplasma* and *Sulfolobus.*

In Lake's classification the eocyte *Sulfolobus* is an excellent candidate for designation of closest modern kin of the original eukaryotic host cell, while *Thermoplasma* is a viable candidate in the midst of Archaebacteria lines. We will focus on these two organisms because significant data about most of the other archaebacteria (or eocytes) is lacking, and the question of which organisms to choose is still unresolved.[1]

In addition we will tackle the problem from the other end. That is, we will examine those eukaryotes that are modern representatives of lineages that branch off at the very base of the eukaryotic

[1] If Lake is correct, then the genera he places in the Eocytes—*Sulfolobus, Pyrodictium, Desulfurococcus, Acidianus,* and *Thermoproteus*—should be a prime focus for analysis.

tree. Some of these most-ancient eukaryotic lines may have retained characteristics of the original host cell.

CHARACTERISTICS OF THE EARLY EUKARYOTIC HOST

THE ORIGINAL host cell must have had some rather unusual features that would serve as precursors for many of the distinctive cell structures we see in extant eukaryotes. These structures include the endoplasmic reticulum (internal membrane system) and its various extensions: the golgi, the nuclear envelope, and membrane-enclosed compartments of all kinds. Mitochondria and plastids are also major and distinctive compartments in eukaryotes, but these are widely understood to have been symbiont add-ons to the original host cell. They are thus the subject of later chapters.

The original eukaryotic host must somehow deliver the initial complexity of the endoplasmic reticulum, golgi, and the nuclear envelope. An important question is, do extant archaebacteria such as *Thermoplasma* or *Sulfolobus* and certain protists on the ancient eukaryotic lineage still bear enough of a resemblance to the original host cells to provide us information?

THERMOPLASMA: RELATIVE OF THE PRE-EUKARYOTIC HOST?

THERMOPLASMA AND other archaebacteria have many unique features, such as distinctive 5S and 16S ribosomal RNA subunits and a lack of peptido-glycans and mureins in their cell walls. They also have an ether-linked lipid with unusual branching, modified transfer RNAs, and a different spectrum of coenzymes. Archaebacteria as a group do, however, have other characteristics that are eukaryote-like. These include occasional presence of introns in the genes for tRNAs, rRNAs, and DNA polymerase (the significance of which will be discussed later), RNA polymerases similar to eukaryotic nuclear polymerases, and the amino acid methionine for initiating translation (Doolittle 1987; Perler et al. 1992).

THE EUKARYOTIC HOST CELL

The focus of this section will be *Thermoplasma acidophilum,* an archaebacterium that completely lacks a cell wall (table 4.1). Found in hot, acidic environments (60°C, pH 1–2) such as coal mine waste piles,[2] these bacteria have features that may conjoin with an early eukaryotic host cell. Some of the very features that were surely selected for survival value in hot, acidic environments might have been fortuitous preadaptations for the acquisition of symbionts by a eukaryotic cell (Margulis and Schwartz 1982). *Thermoplasma* (and another archaebacterium *Methanococcus*) are the only prokaryotes known to have basic histonelike proteins bound to the DNA. Histones are characteristic proteins of eukaryotic DNA, and in this prokaryote may prevent the hot acid denaturation (strand separation) of the DNA (Searcy et al. 1981; Sandman et al. 1990). The calcium-binding protein calmodulin is exclusively eukaryotic, however a likely homologue has been found in *Thermoplasma.* A superoxide dismutase (an oxygen detoxifying enzyme) of *Thermoplasma* also resembles that of eukaryotes (Searcy 1984).

The shape of *Thermoplasma* is quite irregular because it lacks a cell wall. There is, however, evidence of a cytoskeleton, possibly composed of an actinlike protein; this too would be a distinctively eukaryotic characteristic. There are three pieces of evidence for an actinlike protein: (1) *Thermoplasma* growth is inhibited by an anti-actin compound cytochalasin-B; (2) extracts of *Thermoplasma* appear to have an actinlike function, a calcium dependent gel formation alternating with contraction; and (3) filaments of 6 nanometers in diameter were observed by electron microscopy, comparable to 6–10 nm actin filaments of eukaryotes. An actin-myosin system may be of some adaptive advantage in *Thermoplasma* in that these organisms have been observed to flatten themselves on a surface of substrate for the purpose of oxidizing it with a maximum of surface area in contact (Searcy et al. 1981).[3]

This structural system may also, in *Thermoplasma,* have been a preadaption for actin systems used by eukaryotes (Searcy et al. 1981). An actin or actinlike system has an additional value in that phagocy-

[2] Yes, this is an anthropogenic source, but it just happens to be the most convenient place to find *Thermoplasma acidophilum.* Probably other environments such as hot, acidic springs would be appropriate as well.

[3] Mycoplasmas are a group of eubacteria in the gram positive group (not closely related to archaebacteria) that also lack a cell wall and also have actin-like filaments. Perhaps a rudimentary actin-based cytoskeleton is a necessary accompaniment to being wall-less in these two groups.

TABLE 4.1. Similarities between hypothetical early eukaryotic cell with its putative relatives, *Thermoplasma* and *Sulfolobus*.

Hypothetical Early Eukaryote	*Thermoplasma*	*Sulfolobus*
a thermal, acid habitat	yes	yes
actin	maybe an actinlike protein	
intermediate filaments	maybe a laminlike protein	
flexible, wall-less membranes	yes	yes (segmented wall)
steroids		steroidlike molecules? (hypothesized from steroid receptors)
histones associated with DNA	histonelike proteins	
oxygen tolerant cytoplasm, capable of detoxifying oxygen	yes	yes
eukaryotic gene regulation, transcription	some aspects, yes	some aspects, yes
various eukaryotic enzymes and pathways	some such as a calcium-binding protein and a superoxide dismutase	some such as a glutamate dehydrogenase and a sulfur metabolism pathway and a vacuolar ATPase

totic (particle ingesting) activity in eukaryotes depends upon actin. While phagocytosis has not been observed in any extant bacterium, it is hypothesized as a function of the early eukaryotic host cell. *Thermoplasma* may have diverged from an ancestor that had some phagocytotic capabilities or preadaptations.

The presence of actin in *Thermoplasma* is not yet confirmed because some evidence suggests that the cytoskeletal activities may depend on a different group of proteins typically found in eukaryotes, the intermediate filaments (7–11 nm). Filaments of 10 nm in diameter were found in *Thermoplasma*. Although *Thermoplasma* reacts weakly with anti-actin antibody, Hixon and Searcy (1991) conclude that the cytoskeleton may be more similar to a type of intermediate filament that lines eukaryotic nuclei, the nuclear lamins.

Thermoplasma is a fermenter, however it does seem to be aerobic, perhaps owing to a primitive type of respiration that is similar to that of peroxisomes, specialized membrane-bound structures, found in some eukaryotic cells. But *Thermoplasma* uses flavin oxidases to respire, whereas eukaryotic peroxisomes use other oxidases and catalase. *Thermoplasma* lacks a complete electron transport mechanism for either making ATP or pumping out hydrogen ions (as described in chapter 2), but it seems to have a b-type cytochrome (also found in eukaryotic cytoplasm) along with a quinone to reduce oxygen and expel hydrogen ions from the cell. A fermenter producing acids in an acidic environment would necessarily require some type of mechanism for the export of excess hydrogen ions. Thus the use of oxygen in *Thermoplasma* may be attributed entirely to a need for this oxidative mechanism to expel hydrogen ions (Searcy et al.1981; Searcy and Whatley 1982, 1984).

SULFOLOBUS: DISTANT RELATIVE OF THE PRE-EUKARYOTIC HOST?

SULFOLOBUS HAS a similar habitat to that of *Thermoplasma:* hot (70°C) and acidic. It may be isolated from hot acidic springs, specifically from the sulfur particles to which its flat irregular cells are adsorbed (Masover and Hayflick 1981; Vitaya and Toda 1991). *Sulfolobus* is a heterotrophic respirer. A Krebs cycle enzyme (succinate dehydrogenase) has been isolated in this organism (Moll and Shaefer 1991) as well as a cytochrome b that appears to be at the end of an electron transport sequence (Becker and Shaefer 1991). It also seems to be a chemoautotroph, capable of using sulfur compounds as a source of reducing power and energy to fix carbon dioxide for manufacturing sugars (Nixon and Norris 1992). *Sulfolobus* has several proteins in common with eukaryotes but not with eubacteria; an example is a type of glutamate dehydrogenase, an enzyme (Schenkinger et al. 1991). Some aspects of sulfur metabolism, such as a transfer of sulfur from methionine to cysteine, are eukaryote-like (Zhore and White 1991). *Sulfolobus* has a proton-pumping ATPase of the type found in eukaryotic vacuoles (Gogarten et al. 1989). (See table 4.1.)

Some aspects of gene transcription in *Sulfolobus* are similar to those of eukaryotes. These include a subunit of RNA polymerase—an enzyme that makes mRNA from DNA—(Klenk et al. 1992) and a particular regulatory sequence for the RNA polymerase—the TATA box sequence, an area of the DNA rich in thymine and adenine—(Reiter et al. 1990). Most striking is evidence that *Sulfolobus* may have steroids in its membranes, a characteristic usually considered distinctly eukaryotic. This evidence is based on the presence of steroid receptors. The steroids themselves have not been found yet (Muskhelishvili et al. 1990). In *Sulfolobus* steroids may act as "high performance additives," maintaining membranes in near-boiling temperatures, or as "antifreeze" in cooler temperatures.

CHARACTERISTICS OF THE EARLY EUKARYOTE

KEEPING IN mind that *Sulfolobus* like *Thermoplasma* cannot be a direct ancestor of eukaryotes (they both have had their own long evolution in hot, acidic environments) the several characteristics just discussed and summarized in table 4.1 do suggest that the early eukaryote not only was acidophilic and thermophilic but that it also had some elements of the Krebs cycle and perhaps even electron transport—both of which functions are now performed in the mitochondria of eukaryotic cells. Possibly the initial selective impetus for acquiring mitochondria was something other than primary metabolism, as will be suggested in chapter 6. That is, disposing of acidic wastes may have been an important part of the early symbioses, assuming that early eukaryotes were also acid-dwelling thermophiles.

The histones and steroids presumed to have been characteristic of the early eukaryotes would have been logical preadaptations for thermophiles in that these molecules, respectively, protect DNA and membranes from hot or acidic conditions. Later these molecules would confer two of the major eukaryotic characteristics: a stability of the DNA in chromosomes (thus leading to mitosis and meiosis) and an ability of cellular membranes to phagocytize.

THE ACQUISITION OF SYMBIONTS: A MAJOR TRANSFORMATION

IT IS a loss to science that those single-celled organisms that made the transition from pre-eukaryote to eukaryote to symbiont-bearing eukaryote have not left much of a mark in the fossil record. For if they had, the acquisitions of symbionts might have appeared as dramatically transforming events. The newly formed symbiotic cells might have had the appearance of arising de novo, so different would they be from their recent, solitary ancestors.

The later evolution of photosynthetic eukaryotes does, however, appear as a sudden event in the fossil record—preserved thanks to the tendency of photosynthesizers to leave fossils (chapter 1). Extant symbiotic associations abound with examples of rapid transformations which may be so dramatic (including physiology, morphology, and behavior) as to be considered speciation events (Dyer 1989a; Margulis 1991; Bermudes and Back 1991). For example, the great diversity of cellulose-eating animals—wood roaches, termites, ship worms, cows, horses, elephants, giraffes, rabbits, sauropods (herbivorous dinosaurs), hoatzins (a South American bird), and naked mole rats—all have cellulose-digesting symbionts. These are usually contained within specially evolved sections of the hosts' digestive systems. The rumen (fermentation tank) of the cow accounts for about a tenth of the bulk of the animal (about a hundred liters) and gives it a distinctively barrel-shaped form (figure 4.1). Cows and other cellulose eaters) tend to herd or live in colonies, one function of which is to pass symbionts on to offspring that are born with sterile digestive systems. Thus some of the major characteristics of cows as a species—size, shape, choice of food, the distinctive digestive system, a tendency to herd—are direct consequences of their symbiosis with fermenting bacteria. In fact the "sudden" appearance of a large, herbivore in the fossil record, with the bone structure to support a multi-liter fermentation tank, would strongly suggest a corresponding "sudden" acquisition of symbionts. This is the reasoning behind the hypothesis that sauropod dinosaurs must have been filled with cellulose digesting symbionts (Bakker 1986).

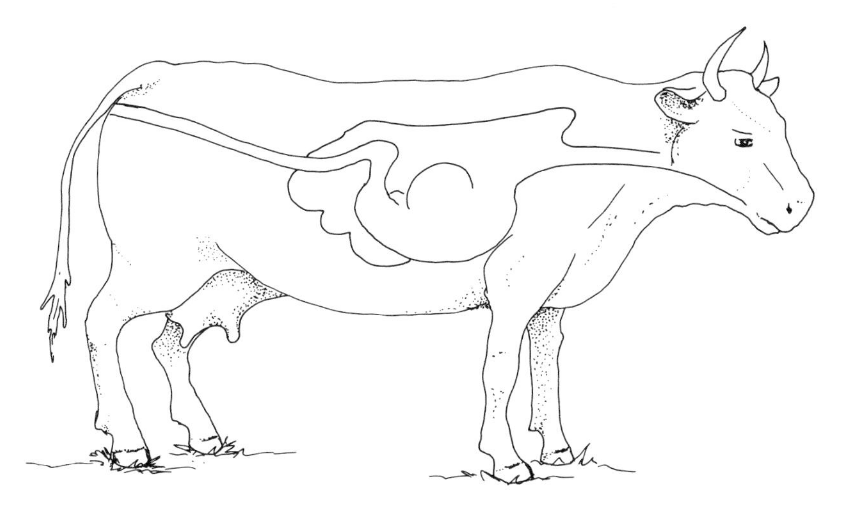

FIGURE 4.1. The modified stomach of the cow forms about a tenth of its body size and is one of the cow's characteristics defined by its symbionts. Drawing by C. Nichols.

HOW FAST IS FAST?

THROUGHOUT THIS book, when we have described an event as happening quickly, we have usually added or implied a qualification that our reference scale is geologic. In geological time a million years is fast, and tens and hundreds of thousands of years are so fast as to be like the blink of an eye in the fossil record; often sediments are dated with the stipulation that the date is accurate within hundreds of thousands and even millions of years. Symbiotic events such as the ones by which eukaryotic cells were assembled most likely occurred so fast that they registered not even a geological blink in the fossil record.

One of the most compelling examples of the rapid establishment of a symbiosis took place in the laboratory of Kwang Jeon (1991). The now-legendary work began in 1966 when Jeon's cultures of *Amoeba proteus* became infected with rod-shaped bacteria, up to 150,000 bacteria per cell. Many amoebae died but a few survived the infection and were maintained in culture. By 1971, just five years later, Jeon had a strain of amoebae that had not only survived the bacterial infection but had evolved a dependence on the invaders.

How did this remarkable transition occur, from pathogen to symbiont? And what is the nature of the host's dependence? Jeon conducted a series of clever experiments, still ongoing, to understand the sequence of events by which the symbionts evolved. Jeon was able to repeat the initial virulent infection with a new uninfected batch of amoebae; he discovered that the transition occurred in about two hundred amoeba generations (cell divisions). At the end of the transition the number of bacteria per cell had been reduced from a high of 150,000 to 42,000, suggesting that the reduction in virulence was due in part to a reduced number of bacteria. The initial entry of the pathogen/symbiont bacteria into the host amoebae probably occurred through phagocytotic ingestion by the amoebae. The bacteria remained undigested, enclosed within vesicles; there they were resistant to lysosomal (digestive) enzymes and may also have prevented fusion of the vesicles with lysosomes (structures in which cellular digestion occurs).

Jeon performed a series of nuclear transplantation experiments to determine whether those amoebae that had become dependent on the bacteria had undergone any nuclear changes. Nuclei extracted from infected amoebae and inserted into "normal" amoebae in which the original nuclei had been removed yielded nonviable amoebae. That is, there seemed to be something lacking in these nuclei, some change or loss that occurred during the infection such that these nuclei could no longer code for (or regulate) a gene product (or products) for "normal" amoebae. Interestingly, these normal amoebae with deficient nuclei could be revived with an injection of bacteria. This demonstrated that the bacteria must be supplying "something" to the host, something no longer coded for or regulated by the nucleus.

Jeon found that growing the infected amoebae at elevated temperatures or in antibodies "cured" the bacterial infection, but killed the dependent amoebae. What exactly the bacteria provide to the amoebae has been difficult to determine. So far, the bacteria cannot be cultured free of their hosts, and the specific function of a polypeptide that they apparently provide to the host has not been discovered. The amazing rapidity of the establishment of this symbiosis, about two hundred amoebae generations, strongly suggests that the appearance of the first taxa of complex cells may also have happened in a flash, geologically speaking. While we are not insisting that mere decades were all that was needed to make eukaryotic cells, it is conceivable that tens or hundreds of thousands of years might have been sufficient to establish the initial steps. The fossil record has probably not retained any such "moment" in eukaryotic evolution but rather shows us the consequences, after the fact.

We hypothesize (with perhaps no chance of seeing fossil evidence) that the eukaryotic transformation was a rapid and dramatic series of symbiont acquisitions. We expect that those few extant eukaryotic groups earmarked as "most primitive" are far removed from their early eukaryotic ancestors, not only because of more than two billion years of subsequent evolution but perhaps more importantly because of the magnitude of the transformation from nonsymbiotic to symbiotic. Nevertheless, we present in the following section the primitive eukaryotes, in hopes that there may still be some provocative hints of the ancestry.

MICROSPORANS AND DIPLOMONADS: DESCENDENTS OF THE EARLIEST EUKARYOTES?

THE FIRST eukaryotes were founding members of the kingdom Protoctista, a diverse group which includes protozoans, algae, motile fungi, and a variety of obscure eukaryotes. The protoctists then gave rise to the other three eukaryotic kingdoms: plant, animal, and fungi. Owing to one or more symbiotic events, all eukaryotes are anaerobic heterotrophs or aerobic heterotrophs or photosynthesizers (figure 4.2). It is the enormous and polyphyletic diversity of the protoctists that enables us to speculate about the nature of the earliest eukaryote and of the subsequent events leading to more complex eukaryotes. We will follow the classification system of Margulis et al. (1989), which recognizes thirty-six phyla of protoctists, but as modified by data from Sogin 1991 (table 4.2). It is a classification system that attempts to deal with the multiple prokaryotic ancestry of the protoctists, which comprises not only a branching tree but also an anastamosing tree because of the symbiotic unions that occurred. Future modifications of the system will have to include the ever-increasing body of information about protoctist rRNA sequences (e.g., Sogin 1991).

The phylogeny of protoctists is best studied by analysis of rRNA sequences because the sequences of ribosomal RNA subunits are considered to be highly conserved evolutionarily and are ubiquitous in organisms. Therefore, sequence comparisons of various rRNA subunits have been used to determine relatedness of the protoctists.

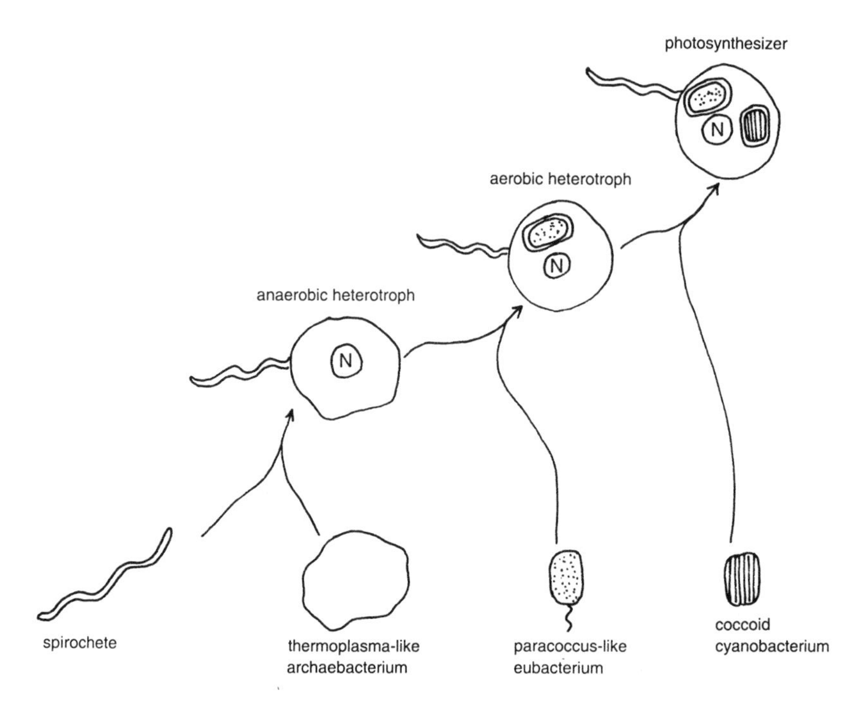

FIGURE 4.2. Symbioses have resulted in three major types of cells in eukaryotes. Drawing after various figures in books by L. Margulis.

Based on 16S rRNA data, the two extant protoctists groups that seem to have descended most directly from the earliest eukaryotes are the diplomonads, a class of Zoomastagina (e.g., *Giardia*), and the microsporans (e.g., *Vairimorpha*) (Sogin et al. 1989). Morphologic evidence is also suggestive of ancient roots: neither of these groups has mitochondria or peroxisomes, and both have microtubular structures.

There is no obvious direct descendent of a protoctist lacking microtubule systems as a primitive feature among any of the thirty-six extant phyla of protoctists. This suggests that motility organelles and various associated microtubular structures evolved or were acquired earlier than mitochondria and plastids. The considerable variation in microtubule structures across the protoctist phyla represents, most likely, extensive adaptive radiation.

Diplomonads such as *Giardia,* an intestinal parasite of mammals and the bane of campers, may be either single cells or mirror-image doubles. They have several motility organelles and undertake the

TABLE 4.2. The protoctist branches in approximate order of antiquity.

		Mitochondria	*Plastids*
Ancient (based upon sequence data)	1. Diplomonads *(Giardia)*	no	no
	2. Microsporidians *(Vairimorpha)*	no	no
	3. Parabasalids *(Trichomonas)*	no* (2° loss?)	no
	4. Euglenoids and Kinetoplastids	yes	some
	5. Some amoebae (e.g. *Naeglaria, Entamoebae)*	some (2° loss?)	no
	6. Cellular slime molds *(Dictyostelium)*	yes	no
	7. Rhodophytes (Red Algae)	yes	yes
	8. Multiple, near simultaneous branches *in no particular order*		
	a. Dinoflagellates, ciliates, apicomplexans	yes	some
	b. Diatoms, Brown and Golden Algae,† Oomycetes	yes	yes
	c. Fungi, Chytrids	yes	no
	d. Animals, Choanoflagellates	yes	no
	e. Plants, Chlorophytes	yes	yes
Recent	f. Haptophytes		

SOURCE: Margulis et al. 1989, with modifications from Sogin 1991.
NOTE: protoctist groups that do not yet have a place on the tree are Pelomyxans, and some Zoomastigina such as Retortomonads, Pyrsonymphids (all lacking mitochondria and plastids) as well as various photosynthetic groups, Chloroarachnids, Prymnesiophytes, Eustigmatophytes, and also some heterotrophic groups such as Granuloreticulosa (foraminifera), plasmodiophoromycota, hyphochytridiomycota, actinopoda, haplosporidea, paramyxea, myxozoa, xenophyophora.
*This group has an organelle, the hydrogenosome, which is either a degenerate mitochondrion or an organelle of separate symbiotic origin (more in chapter 6).
†Golden Algae include chrysophytes and xanthophytes.

kind of cell division (mitosis) typical of eukaryotic cells (Vickerman 1989).

Microsporans are all parasitic, mostly of invertebrates but sometimes of other protoctists. They have none of the membrane-bound packages (lysosomes) typical of eukaryotes, and they lack the golgi apparatus and motility organelles. The spindles used in the mitotic division of Microsporans lack microtubule structures found in many but not all eukaryotes. However, some microsporans have an atypical meiotic sexuality (Canning 1989, 1988).

Diplomonads and microsporans both have some very specific and unique adaptations to a parasitic life cycle. For example, *Giardia* has suckers with which it attaches to the host intestine, and the microsporans inject an infective cell into a host cell via a polar tube. If these two groups are direct descendants of early eukaryotes, it may be that subsequent parasitism evolved as a particularly advantageous mode of nutrition for cells lacking mitochondria. It should be kept in mind that a parasitic (one form of symbiotic) relationship exerts a powerful selection pressure on both host and symbiont and that many features of diplomonads and microsporans may be specializations and not primitive characteristics. Also, it is possible that both groups may have unusually high rates of evolution, resulting in an inaccurate rRNA tree (Siddall et al. 1992). It is difficult therefore to evaluate the relevance of a single piece of information, such as the fact that microsporans lack lysosomes; is it a recently derived specialization or a primitive characteristic? These two mostly parasitic groups may or may not yield reliable evidence about the structures and functions of the original pre-eukaryotic cell. The only conclusion that can be made with some confidence from these protoctist groups is that microtubular (tubulin) structures, as well as other cytoskeletal elements, preceded both mitochondria and plastids.

THE NEED FOR FURTHER WORK ON PROTOCTIST rRNA

SEQUENCING AND analysis of ribosomal RNAs is now an area of active research, with new findings rapidly making obsolete any previously published work. Our conclusions here, although based on the best available evidence, should be read with this caveat in mind.

At least three groups of protoctists lacking mitochondria, besides the diplomonads and the microsporans, should be examined to determine their position on the rRNA family tree. These include two classes of the phylum Zoomastigina—the retortamonads (Brugerolle and Mignot 1989) and the pyrsonymphids (Dyer 1989b; see figure 4.3)—and the phylum Karyoblastea (Whatley and Chapman-Anderson 1989). The retortamonads and pyrsonymphids are obligate symbionts of animals and both have motility organelles but no mitochondria. The karyoblasteans are giant (0.1–5.0 mm) free-living amoebae without mitochondria, but loaded with symbiotic bacteria; they seem to have a microtubule cytoskeleton, although spindles and other microtubule organizers have never been identified in these amoebae (Whatley and Chapman-Anderson 1989; Whatley and Whatley 1983). An analysis of ribosomal RNA, not yet performed, should indicate whether or not these organisms too are direct descendants of the earliest eukaryotes, or whether their lack of mitochondria represents secondary loss.

It is a secondary loss that probably best accounts for the lack of mitochondria in several other protoctists. The ciliates are a cohesive and rather advanced group, some members of which lack mitochondria and are adapted to live in deep, anaerobic sediments (Fenchel et

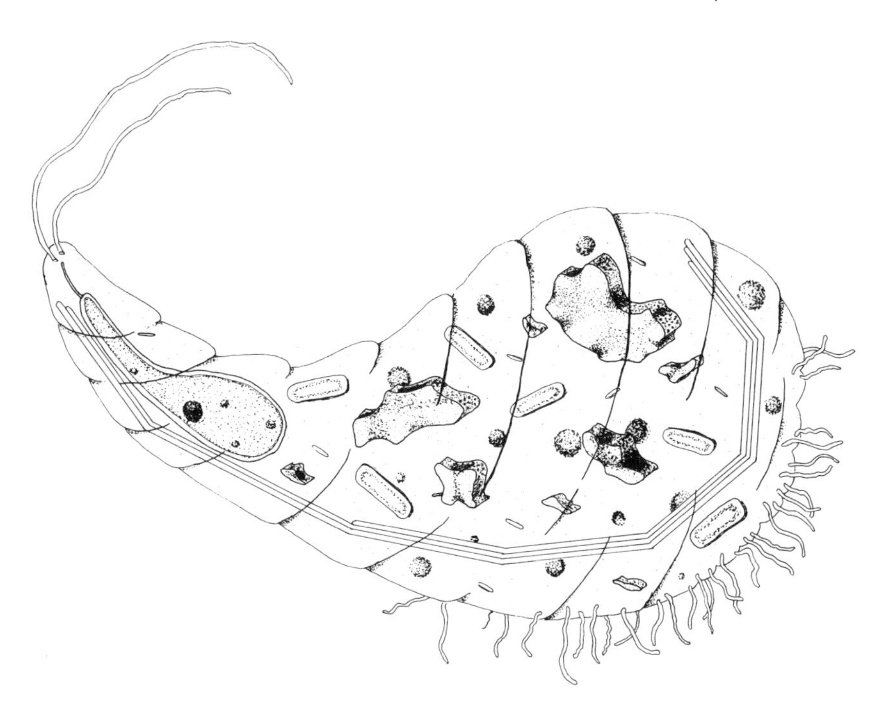

FIGURE 4.3. *Pyrsonympha* is one of the presumed early eukaryotes (lacking mitochondria) that should have its rRNA sequences studied. About 100 micrometers long, it is found in the hindguts of termites. Drawing by C. Nichols.

al. 1977; Dyer 1989c). The rhizopodan amoebae may or may not be such a cohesive group, depending upon the outcome of further rRNA studies. One rhizopodan, *Entamoeba histolytica,* an intestinal parasite, stands out as lacking mitochondria, but this appears to be a secondary loss. The parabasalids (a class of Zoomastigina) are uniformly devoid of mitochondria, and they are all symbionts of animals. One parabasalid, *Trichomonas vaginalis,* a parasite of humans, has rRNA that suggests its group closely follows the branching of Diplomonads and microsporans on the ancestral tree (Sogin et al. 1989; Sogin 1991). It is not clear in this group whether the lack of mitochondria is primary or secondary. Alternatively, parabasalids may not have lost their mitochondria as they still have mitochondria-like structures, hydrogenosomes, that may be derived from mitochondria (see chapter 6).

At least fifteen phyla of protoctists have photosynthetic plastids, and all of them have mitochondria as well. This suggests that both organelles were acquired almost simultaneously in these groups and/or that there is some adaptive advantage in having plastids with mitochondria. Indeed, mitochondria seem to be an integral part of a glyoxylate cycle with plastids (described in chapter 7). Ribosomal RNA data suggest that some of the first eukaryotes to acquire mitochondria are direct ancestors of both Euglenoids and Trypanosomes. Most of the euglenoids also have plastids. Important questions are: Were mitochondria acquired more than once in protoctists, that is, are they polyphyletic? And were plastids acquired more than once?

Notice in table 4.2 that groups branching off later than the diplomonads and microsporans either have mitochondria or appear to have secondarily lost mitochondria (perhaps on making a transition into an anaerobic environment). Parabasalids, because of their branching at what looks like a pivotal point, require considerable further analysis; either they have never had mitochondria or they have secondarily lost mitochondria or the hydrogenosome, another organelle in these organisms, is a degenerate mitochondrion (see chapter 6). No evidence suggests as yet that mitochondria arose more than once, although new developments could change that supposition.

The plastids are a somewhat different story in that there are at least five cohesive and separate branches of plastidic eukaryotes (table 4.2). Euglenoids and Rhodophytes seem to have acquired photosynthetic symbionts independently, as well as certain distinctive features of their plastids, thus reflecting distinct lineages on the family tree. The other plastidic branches present more of a problem in that they appear to have diverged nearly simultaneously. Whether or not the abundant green plastids, the plastids of dinoflagellates, and the plastids shared by diatoms, brown algae, and golden algae are truly separate acquisitions is completely unknown and deserves thorough investigation with rRNA analysis.

ENDOPLASMIC RETICULUM, ENDOCYTOSIS, AND EXOCYTOSIS

EUKARYOTIC CYTOPLASM is full of membranous structures and cytoskeletal elements, which are apparently not of symbiotic origin. These must have been either part of the original composition of the early eukaryotic host or the result of subsequent evolution in the earliest, successful lineage of eukaryote.

One such structure is the endoplasmic reticulum (E.R.). It is a three-dimensional labyrinth of folded membranes, some of which may be studded with ribosomes (figure 4.4). There is a sidedness to

FIGURE 4.4. The labyrinthine membranes of the endoplasmic reticulum fill the eukaryotic cell and also are part of the nuclear envelope. Drawing by C. Nichols.

the E.R. membranes, in that the inner surfaces border a convoluted lumen space and are relatively smooth (figure 4.5). The outer surfaces, often embedded with ribosomes, are apparently confluent with the cytoplasm of the cell. In cross section it can be seen that the lumen spaces alternate with the cytoplasm (cytosolic spaces). Polypeptides tend to be transported across E.R. membranes from the cytostolic side to the lumen side, pushed through as they are translated by the attached ribosomes. The mechanism, called the signal hypothesis, by which specific polypeptides are inserted into membranes or pushed through at specific sites is not unique to eukaryotes. However in prokaryotes, polypeptides are usually pushed through from the inside to the outside of the cell.

Topologically the two systems, prokaryotic and eukaryotic, might be similar in that endoplasmic reticulum probably evolved from convolutions of the plasma membrane that budded off inside the cell to enclose lumens (figure 4.6). Thus what is now the surface facing into the lumen of the E.R. was once functionally the outside surface of the prokaryotic ancestor. This is supported by the fact that the lumen spaces of E.R. often bud off small vesicles containing specific macromolecules, which are transported to the plasma membrane. Then the E.R. membrane and plasma membrane fuse in such a way that the lumen space is turned inside out, dumping its contents into the environment (exocytosis). At this point the inner lumen face of the E.R. is continuous with the outer surface of the plasma membrane (figure 4.6).

Endocytosis is the reverse of exocytosis and enables "new" endoplasmic reticulum to be formed from an invagination of the plasma membrane. This is a means by which cells can take up macromolecules and particles from the environment (figure 4.6). Both exocytosis and endocytosis are unknown in prokaryotes. They may have been present in the pre-eukaryotic host cell. Endocytosis in particular would have been a mechanism by which endoplasmic reticulum might have been formed and by which symbionts might have been acquired. The likely presence of steroids and actin in the membranes of early eukaryotes to assist their survival in hot, acidic environments might have been preadaptations leading to exo- and endocytosis.

Although all known prokaryotes lack endocytosis and exocytosis, they certainly do not lack convoluted internal membrane systems (Shivley 1974). These presumably came about from infoldings of the plasma membrane. Unfortunately, *E. coli* is too often used as the

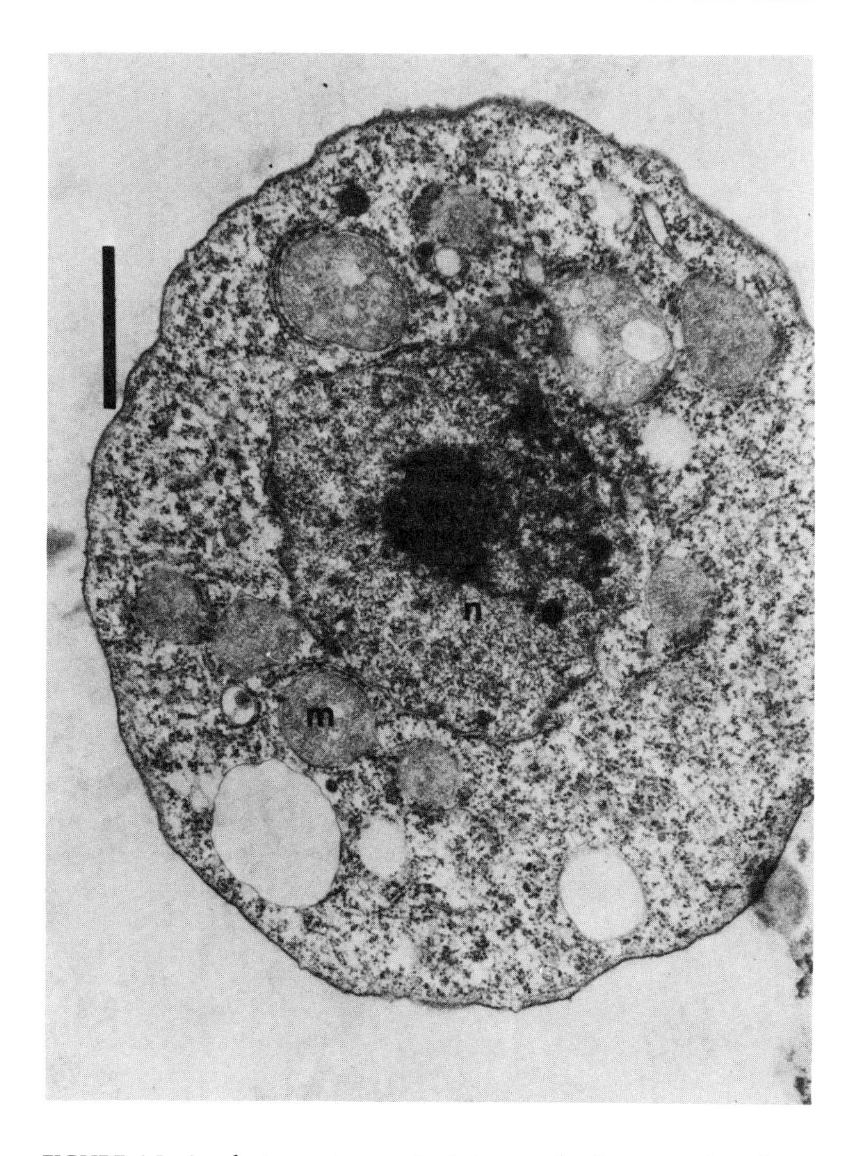

FIGURE 4.5. An electron micrograph of the amoeba *Paratetramitus* shows that the endoplasmic reticulum can actually appear (in cross section) to be like small broken pieces of membrane, some studded with ribosomes, throughout the cytoplasm (n indicates nucleus; m indicates mitochondria; the bar scale is 1 micrometer). Photograph by B. Dyer.

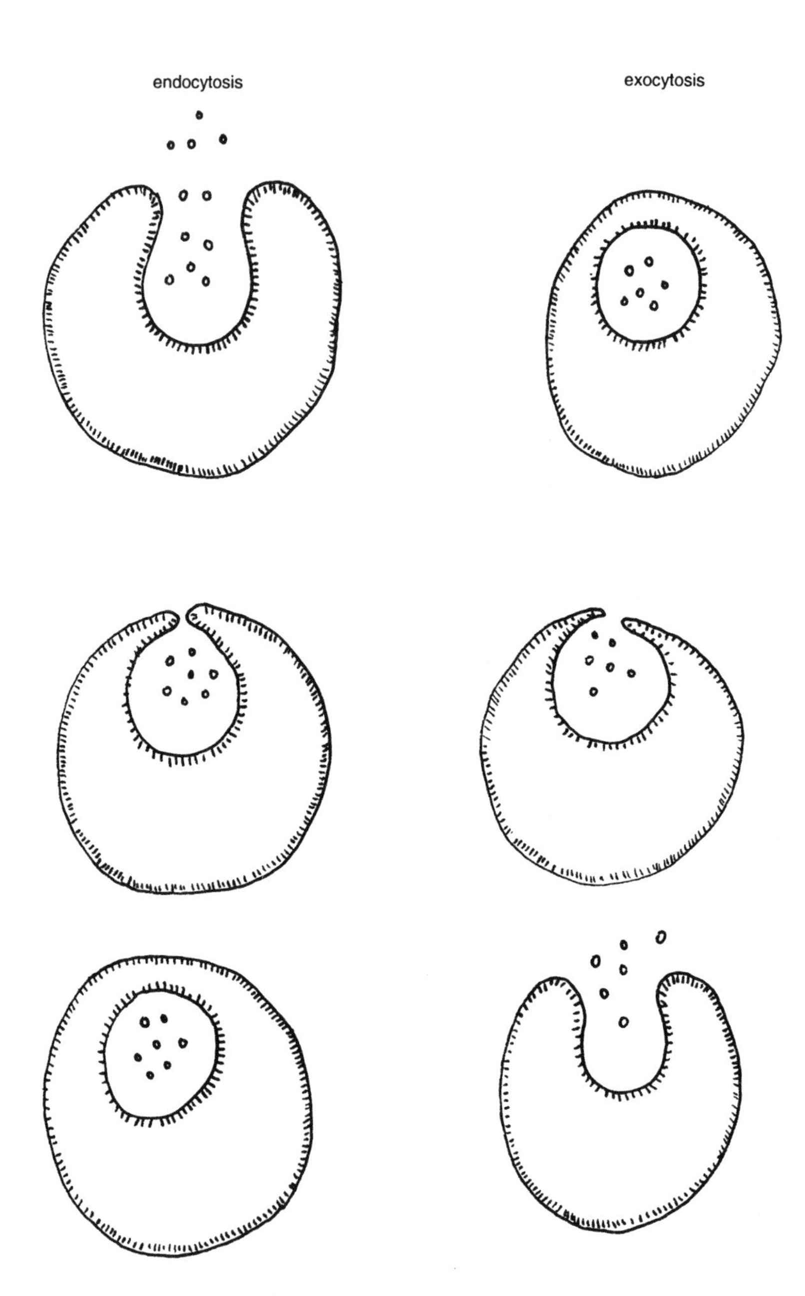

FIGURE 4.6. During endocytosis new endoplasmic reticulum is formed by an invagination of the cell membrane. When eukaryotes form small vesicles (exocytosis), the membranes that were facing inward to the lumen space become continuous with the outer membrane.

standard for prokaryotes; this bacterium can give a misconception that bacterial membranes are always rigidly held in place by a cell wall and that bacterial cells are rather simple inside (with no internal membranes). Actually, *E. coli* (the white rat of medical microbiologists) is of a rather specific group of enterobacteria, found easily in mammalian digestive systems.

If we look at the whole diversity of bacteria (e.g., through *The Prokaryotes,* edited by Starr et al. 1981) we see many important groups with internal membrane systems. These include the thylakoid membranes of cyanobacteria (figure 4.7), lamellae of certain chemoautotrophic and photosynthetic bacteria, membrane vesicles of photosynthetic bacteria, and mesosomes in many bacterial groups (including *E. coli*). Several of the Archaebacteria are known to have especially flexible membranes, such as *Sulfolobus* which often has an irregular lobed appearance for it is nearly wall-less. The feeble wall that does exist lacks peptidoglycan and is probably composed entirely of protein in a protein-lipid complex. Furthermore it may contain steroids that would enhance flexibility.

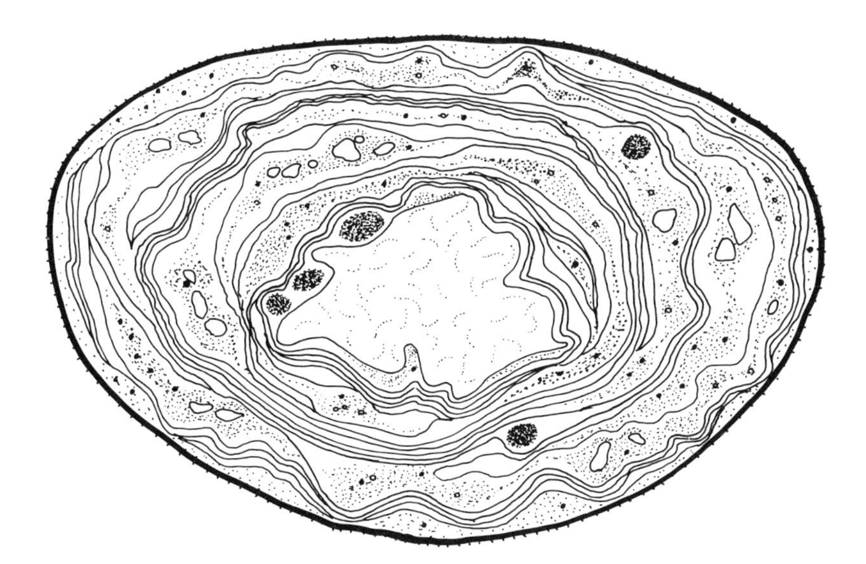

FIGURE 4.7. Cyanobacteria have an internal membrane system, the thylakoids, on which some of the reactions of photosynthesis occur. Drawing by C. Nichols.

It is conceivable that mechanisms for folding membranes inward to form compartmentalized interiors were and still are present in many prokaryotic groups, including the archaebacteria. These mechanisms, whatever they may be, are probably the preadaptations that gave rise to endoplasmic reticulum (as a unique eukaryotic system), as well as to exocytosis and endocytosis. While these mechanisms are not unique to archaebacteria, this group does have the additional feature of highly flexible membranes (and modified or nonexistent cell walls), which makes it a good candidate for precurser in the evolution of endocytosis and exocytosis.

Why then are endocytosis and exocytosis never seen in prokaryotes? There are two possibilities. First, perhaps some prokaryotes are in fact capable of endo- and exocytosis; the specific taxa may have simply not yet been studied. This is a real possibility, considering the unusual environments and difficult culture conditions for most archaebacteria. Certain observations about cytoskeleton-like structures in the archaebacteria do not preclude this possibility. Second, it is possible that endo- and exocytosis, if and when they evolved in some ancient pre-eukaryotic line, were such an instant success—enabling the owner to gobble large molecules and even other organisms—that members of the fortunate lineage all irrevocably became the eukaryotes. Circular reasoning may be unavoidable if this is the case.

THE NUCLEAR ENVELOPE

THE NUCLEAR envelope is topologically continuous with the endoplasmic reticulum (E.R.) and probably arose directly from E.R. membranes (figure 4.4). In some ways, the nuclear space may be thought of as an area defined by highly modified endoplasmic reticulum. There is, however, no mistaking the nuclear envelope for ordinary E.R.

The nucleus is often the most visible structure in a eukaryotic cell that can be observed with the light microscope.[4] The chromo-

[4]The nucleus (containing most of the DNA) is a primary indentifying characteristic of all eukaryotes. It was first observed in a ciliate by A. van Leeuwenhoek (1623–1723). "These animals had a pretty structure, for the round circumference of their bodies seemed to be made up of ten or twelve bright, round pellets while

somes inside often stain uniquely, making the nucleus even more visible. The activities of the nucleus also separate it from those of the E.R. In many organisms the envelope disintegrates during mitosis and meiosis and is then reconstructed afterward. The nucleus specifically maintains an internal environment for DNA and its associated proteins and for newly made RNAs of all types. Only certain molecules generated in the nucleus, such as RNAs, are allowed to leave through the nuclear pores; even fewer are allowed to enter, mostly the various polypeptides that are synthesized in the cytoplasm and then shunted to the nucleus for specific functions. These include DNA polymerase, RNA polymerase, and the proteins associated with the DNA and the inside walls of the nucleus.

Although there is considerable variation in eukaryotes in respect to number, size, and shape of nuclei as well as whether or not the envelope breaks down in cell division, all have a well-defined nucleus (or more than one). In fact, the nucleus is the universal defining characteristic of eukaryotes. (According to Fuerst and Webb 1991, the nucleus-like structure found in the eubacterium *Gemmata* is strictly a case of convergent evolution, and is very exceptional in a prokaryote.) The lack of an obvious intermediate structure in an extant eukaryote group does pose a problem in tracing the evolution of the nucleus. This may be yet another example of an adaptation (in this case a modification of the endoplasmic reticulum) that was such a resounding success that any eukaryote without one was quickly outcompeted.

The nucleus must have evolved quite early in the history of eukaryotes or perhaps even in the pre-eukaryotic prokaryote, given its universality. There must indeed be some sort of selection pressure for keeping genetic material together in one or more specific areas of the cell. This is often observed in prokaryotes, which, while they have no nucleus, do have distinct features called nucleoids. The increasing levels of oxygen in the atmosphere may also have selected for the nucleus as a mechanism for maintaining a relatively oxygen-free environment for the genes, which are especially sensitive to oxidation (Gupta et al. 1990; Kvam et al. 1990).

---

in the middle of them there seemed to be a little dark spot somewhat bigger than the pellets" (Dobell, 1958).

MICROBODIES (PEROXISOMES): WERE THEY SYMBIONTS TOO?

UNLIKE THE numerous other membrane-bound compartments of the cell—the vesicles, lysosomes, vacuoles, and golgi structures—the microbodies (also called peroxisomes) are apparently not derived of endoplasmic reticulum. Microbodies may be found in close proximity to E.R., but their formation seems to be independent. Some microbodies contain peroxidases, catalases, and sometimes oxidases. In plant photosynthetic cells, microbodies called glyoxysomes may have the additional function of photorespiration (the glyoxylate cycle, an alternative pathway of the Calvin cycle). In a parasitic group of protoctists, the trypanosomes, microbodies called glycosomes participate in a variation of the Krebs cycle by which fatty acids may be metabolized.

Microbodies (peroxisomes) either evolved in eukaryotes endogenously or were acquired as symbionts. If they were acquired, the symbiogenesis occurred at some point during or after the evolution of mitochondria, because some presumably ancient protoctist groups lacking mitochondria—the microsporans, diplomonads, pelomyxans, and retortamonads—also lack microbodies (Cavalier-Smith 1987a, 1987b). Confusing the issue, *Entamoeba,* which lacks both mitochondria (probably a secondary loss) and microbodies, appears to be on a more recent branch than the ancient groups just mentioned; if microbodies are of symbiotic origin, *Entamoeba* may have secondarily lost them along with mitochondria. The evidence is thus somewhat contradictory for either the endogenous origin or the symbiosis hypothesis.

One scenario of the endogenous origin hypothesis is that peroxisome-like functions were already present in the pre-eukaryotic host and later became enclosed in a single membrane. Although the anaerobic glycolytic (fermentative) pathway dominates the metabolism of the cytoplasm of all eukaryotic heterotrophs, the cytoplasm is not strictly anaerobic, as evidenced by the activities of peroxisomes. The pre-eukaryotic host was probably not strictly anaerobic either. Searcy and Whatley (1982, 1984) predicted that they would find the metabolism of *Thermoplasma* to be similar to that of the modern eukaryotic cytoplasm. In addition to glycolysis, they suggested that respiratory activities similar to those of peroxisomes might be present. Glycolysis does in fact account for most of the glucose metabolism of *Thermoplasma. Thermoplasma* also appears to have a partially functioning Krebs cycle that excretes acetic acid as a waste product and that does not link up to an electron transport system. Substrate-level phosphorylation appears to be the only method by which *Thermoplasma* produces ATP. Searcy and Whatley (1982) also found evidence for peroxisome-like functions in *Thermoplasma.*

Alternatively, eukaryotic microbodies may have originated through symbiosis. Microbodies appear to go through a budding type of division to produce more microbodies. They do not seem to form de novo in the cytoplasm (Opperdoes and Michel 1989; Tabak and Distel 1989). This is symbiont-like behavior, but there is no trace of a genome in microbodies. Lack of a genome remnant implies either that microbodies are not symbionts or that they horizontally transferred and/or lost every one of their genes.

The single membrane that encases each microbody is also somewhat atypical for symbionts, which are usually bounded in double membranes. It is, of course, conceivable that the microbody-symbionts might have lost the extra membrane, owing to the degradation caused by the act of phagocytosis by the host. The fact that the enzyme pathways within microbodies are complex somewhat supports a symbiotic hypothesis, as a once-independent lifestyle would be the most efficient way to evolve an enclosed complex pathway.

Hydrogenosomes, organelles bounded by double membranes, in some anaerobic protoctists may be highly modified mitochondria; yet the search for DNA inside the hydrogenosomes has yielded ambiguous results (chapter 6). Their situation, a complete loss of an alleged symbiotic genome, could be similar to such a loss in microbodies and, as we will argue, to such a loss in motility organelles. Meanwhile, more evidence is needed to make a conclusion.

THE CYTOSKELETON AND CELL MOTILITY

THE EUKARYOTIC cytoplasm is filled with a network of proteins and membranes composing the cytoskeleton. The two major functions of the cytoskeleton are to maintain cell shape and to confer cell motility both for the entire cell and for the structures inside. One major component of the cytoskeleton is the microtrabecular matrix, a membrane system that is continuous with and derived from the endoplasmic reticulum (Porter and Tucker 1981). Protein components of the microtrabecular matrix include elongated, tubelike microtubules composed of proteins called tubulins.

Microtubules have many important functions, such as maintenance of cell shape, movement of chromosomes during cell division (as spindles), transport of granules and vacuoles in the cell, and movement of motility organelles. All of these microtubule functions require the presence of other specific proteins called microtubule associated proteins (MAPs). MAPs, which probably number in the hundreds of varieties, seem to confer the specific functions on microtubules, enabling the cell to use microtubules in diverse ways.

The other major protein component of the cytoskeleton is actin, which takes the form of microfilaments. Actin also has numerous associated proteins which confer a variety of functions including amoeboid movement, endo- and exocytosis, and contraction of cells (such as that performed in muscular contraction).

Intermediate filaments are a third protein component of the cytoskeleton. These filaments are built of various proteins including keratin, vimentin, and desmin. Intermediate filaments may maintain cell shape, and those that line the inside of the nuclear envelope may help organize the nucleus.

An important question in understanding the origin of eukaryotic cells is how did this complicated and multi-functional system, the cytoskeleton, evolve? What counterparts might be found in prokaryotes?

Very few counterparts to the eukaryotic cytoskeleton have been found in prokaryotes. Although there are many prokaryotes with internal membranes, none have an internal structure as extensive and complex as those of even the simplest eukaryotes. The microtrabecular matrix unique to eukaryotes is contiguous with and probably evolved from another unique feature of eukaryotes, the endoplasmic reticulum, which in turn may have evolved from the internal membranes of a pre-eukaryotic or early eukaryotic host.

Tubulin, actin, and intermediate filaments have been sought in prokaryotes, with inconclusive results. Tubulin is discussed in detail in the chapter on motility organelles, because it is crucial to a hypothesis that all tubulin structures in the cell were acquired by the eukaryotic host when it acquired spirochetes as motility symbionts. The evidence for tubulin or microtubules in spirochetes is inconclusive but suggestive that the symbiogenesis hypothesis may be valid. If indeed symbiotic spirochetes did provide tubulin to their host, they did it very early in the history of eukaryotes, for there appear to be no eukaryotes without microtubule systems. Furthermore, if symbiotic spirochetes supplied the tubulin, then the transfer of the gene for tubulin from spirochete to host must have happened quickly.

Tubulin genes are found only in eukaryotic nuclei. If there is a remnant of a spirochete genome, it has lost most of its original functions. (For details see chapter 8.) It is MAPs (microtubule associated proteins) that confer a diversity of functions onto the cytoskeleton, and so the horizontal acquisition of a tubulin gene alone might have been sufficient to trigger the adaptive radiation of tubulin functions in eukaryotes. Research on MAPs is an active area, but it is not yet known whether any of these distinctively eukaryotic proteins have prokaryotic counterparts. It may be that many of the MAPs that serve as ATPases and GTPases (enzymes that use the energy in ATP and GTP) were preadapted to function with tubulins and arrived in the spirochete or were already present in the host.

Actin seems not to be present in prokaryotes—with the possible exception of an archaebacterium, *Thermoplasma,* and some mycoplasms. The evidence of absence is still, however, preliminary. It consists mainly of observations that *Thermoplasma* has a protein that behaves somewhat like actin and forms filaments of approximately the right size. Thus actin may have arrived as part of the original eukaryotic host cell. Alternatively, the uncharacterized protein of *Thermoplasma* may form an intermediate filament.

INTRONS AND EUKARYOTIC EVOLUTION

INTRONS (I) ARE noncoding sequences of DNA that interrupt gene sequences, exons (E).

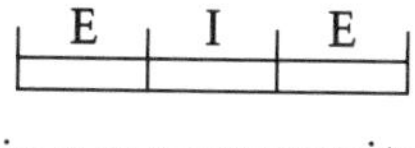

E I E

The presence of introns in a gene necessitates one of several mechanisms to produce a final protein product from that gene. In general, a messenger RNA is transcribed from the gene such that the introns are also transcribed.

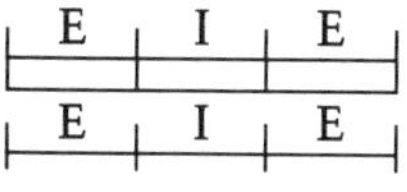

E I E

E I E

Then by a splicing mechanism, the introns are cut out and the exons are joined together to form a mature mRNA ready for translation.

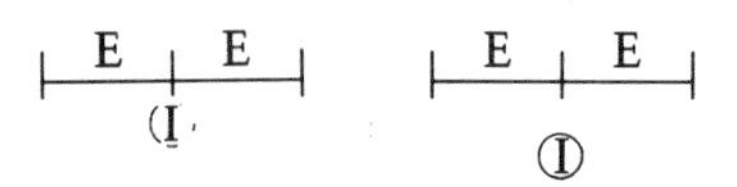

Introns are divided into four groups depending upon their sequence, their usual location in a cell or organism, and their splicing mechanism. Group I is found in a wide variety of organisms and is self-splicing; that is, a part of the mRNA is capable of acting as a splicing enzyme to independently remove the introns. Group I introns appear to be a category of transposon or jumping gene. That is, they are capable of moving about in the genome semi-independently. Group II introns are also self splicing, but they are only found in mitochondria and plastid DNA. Group III introns are found in some eukaryotic nuclei and animal viruses, and are spliced by a ribonucleoprotein complex called a spliceosome. Group III introns are not as autonomous as a type I or II. Group IV introns are found in rRNA and tRNA genes of eukaryotes and archaebacteria, and are spliced by enzymes.

When and how did the four groups of introns arise and how do they fit into the phylogenetic tree of organisms and, in particular, that of eukaryotes? A summary of conclusions is found in table 4.3.

Group I: Wide Distribution; Self Splicing

Palmer and Logsdon (1991) suggest that most Group I introns are recent developments because, despite widespread distribution, they are strikingly absent in many ancient lineages. Group I introns seem to be particularly mobile and independent. Thus they may not be reliable markers in tracing the lineage of organisms. Group I introns may be found in the DNA of fungi mitochondria and plant plastids—both of which are organelles of recent groups of eukaryotes. They have not been found in the organelles of members of the oldest eukaryotic lineage: the protoctists. But Group I introns are found in some protoctist nuclei. The Group I introns found in some bacteriophages (bacterial viruses) are considered by Palmer and Logsdon to be derived from mitochondrial introns. That is, at some point in the history of mitochondria or premitochondrial prokaryotes, bacteriophages acquired some of the mitochondrial sequences that included introns.

A possibly exceptional and distinctive variety of Group I intron is

TABLE 4.3. Conclusions about eukaryotic evolution based on observations about introns.

	Found in	*Spliced by*	*When Evolved*	*Comments*
Type I Intron	mitochondria of plants, fungi some protoctist nuclei some bacteriophages	themselves (rather autonomous type)	recent	may be too mobile and ubiquitous to be good indicators of lineage
Type I Intron	most cyanobacteria most plastids some eubacteria (always in tRNA leu genes)	themselves (but strangely limited to tRNA leu genes)	ancient	evidence for the common ancestry of plastids and cyanobacteria
Type II Intron	mitochondrion and plastids of recent eukaryotes	themselves	recent	confirms lineage of recent and ancient eukaryotic groups
Type III Intron	recent eukaryotic nuclei animal viruses	spliceosomes	recent	Type II's transferred to nucleus, may have become spliceosome-dependent Type III's and organelle genes transferred to the nucleus acquired group III introns from the nucleus
Type IV Intron	eukaryotic nuclei archaebacteria (in rRNA and tRNA genes and DNA polymerase genes)	enzymes	ancient	evidence for relatedness of eukaryotes and archaebacteria

found in most cyanobacteria and eukaryotic plastids that have been examined for such. It is inserted specifically within the leu tRNA genes. It also appears sporadically in the leu tRNA genes of various eubacteria. This exceptional Group I intron may indeed be ancient, dating back to the evolution of cyanobacteria.

Group II: In mitochondria and plastids; self splicing

Palmer and Logsdon (1991) consider Group II introns to be relatively recent because they are found in the mitochondria and plastids of recent eukaryotic groups only: fungi, green algae, and green plants.

Group III: In Eukaryotic Nuclei; Spliced by Spliceosomes

Group III introns follow the family tree of eukaryotes with remarkable fidelity (Palmer and Logdson 1991). Animals, fungi, and plants all have large numbers of them in their nuclei as do animal viruses, which must have acquired them from animal hosts. Fewer and fewer Group III (nuclear) introns are found in earlier protoctist lineages. For example, ciliates, which are recent protoctists, have a significant number of Group III nuclear introns, but ancient groups like diplomonads and microsporans have none at all. Furthermore, whatever genes got transferred from mitochondria or plastids into the host nucleus also acquired Group III introns, at least in the more recent eukaryotes. Group III introns are probably therefore a good example of a late-evolving intron. Group III introns may be Group II introns that were transferred from organelles to nucleus, and which subsequently lost the ability to self splice—thus becoming spliceosome-dependent.

Group IV: In Eukaryotes and Archaebacteria; Spliced by Enzymes

The fact that eukaryotes and archaebacteria share a particular group of introns found exclusively in rRNA and tRNA genes is another piece of evidence of the common ancestry of these groups (Doolittle 1987). Group IV introns must be ancient, and may even be good markers for the eukaryotic and archaebacterial lineages.

THE TWO varieties of intron of obvious ancient lineage, Group I of cyanobacteria and plastids and Group IV of Archaebacteria and

Eukaryotes, are highly specific in their placement within tRNA or rRNA genes. All other introns appear to be more recent. However, alternative interpretations of the data have been proposed. One alternative hypothesis is that most introns would be considered ancient (e.g., Darnell and Doolittle 1986; Belfort 1991). The argument for the lack of introns in most eubacteria (except for those that joined eukaryotes in symbiosis) is that introns have been secondarily lost. Bacterial genomes tend to be small and streamlined; any intron acquired in a eubacterial lineage might be quickly jettisoned in most groups. Those bacteria associated with a host cell might have launched the rejected DNA right into the host nucleus.

The eukaryotes, on the other hand, tolerate enormous, bulky genomes and would be expected to display the greatest number of introns, as indeed they do. In fact the evolution of the nucleus may have been instrumental in this tolerance for extra DNA. The nuclear membrane separates transcription (formation of mRNA) from translation (synthesis of proteins)—two processes that are tightly coupled in bacteria. In eukaryotes this separation might have allowed some crucial extra time for regulatory processes, such as that which splices out introns after transcription.

The argument as to whether introns are ancient or recent may not be resolved quickly. However, for the purposes of this chapter there are two definite conclusions about particular kinds of introns and their significance in the evolution of eukaryotes. First, the Group IV introns conveniently link the lineage of eukaryotes and archaebacteria. Second, it appears that when genes are transferred from the genome of a symbiont (plastid or mitochondria) to the host nucleus they can acquire the Group III introns of their host.

THE READER at this point (having presumably made it through this difficult chapter) might long for some definitive conclusion about those early eukaryotic cells. Alas, none will be presented here as we would not be honest with ourselves as scientists if we tried to make this story sound complete. The unusual prokaryotes, *Thermoplasma* and *Sulfolobus,* and those eukaryotes of ancient lineage, microsporans, diplomonads, and others, are uneasy participants in this chapter. All are extant and have had long evolutionary histories in unique environments. Only some of their characteristics seem to provide clues about the first eukaryotes. A hot, acidic environment (the habitat for the two prokaryotes) may have selected for stable DNA structures and flexible cell membranes, characteristics that would later be useful in the first eukaryotic cells. The transformation, via symbiosis, might have been both rapid (on a geological scale) and dramatic. However those early events are unlikely to have a clear fossil record. Only evidence from extant symbioses suggests how such transformations might have come about. The cell structures of the eukaryotes on the ancient lineage provide clues about the order of events: motility organelles and nucleus (simultaneously?), followed by mitochondria and then plastids. However the eukaryotes used most often as examples may not be the best; more information is needed about other eukaryotic groups such as the Karyoblastea and Pyrsonymphida.

In fact, rather than present any definitive conclusions about evolution of eukaryotic nuclei and other cytoplasmic components, we prefer to caution the reader against readily accepting any simplified story about early eukaryotes. This chapter on the evolution of the nucleus and cytoplasm is an example of a small branch of science at a frontier. Data and observations will slowly accumulate. New interpretations may come about. Watch for developments in the future.

V

The Evolution of Eukaryotic (Meiotic) Sex

BIOLOGISTS KNOW a great deal about the sexuality of animals (and especially of vertebrates). The general characteristics of sex in vertebrates are often considered to be the normal ones with which all other eukaryotes are compared. While a vertebrate model has been and continues to be a very useful framework for research, it obscures the enormous variety of sexual functions found in fungi, plants, and especially protoctists. Mammals, in particular, are peculiarly limited in that they cannot reproduce by budding or dividing and are confined to reproduction linked with sex. What is by far the most common practice in both prokaryotes and eukaryotes is reproduction without any sexuality at all. Because protoctists are highly variable with respect to sexuality and reproduction, and because their lineage is the most ancient of all extant eukaryotes, they will be the focus of this chapter.

It all begins with the rather promiscuous mixing of DNA typical of bacteria. Approximately two billion years ago, some of this earliest form of sexuality was channeled into a new and emerging set of mechanisms that would become meiotic (eukaryotic) sex. How did it happen?

The clues reside in the diversity of protoctists, major lineages of which seem to have branched off at various points during the evolution of meiotic sex. Tracing the evolution of meiotic sex is by no means a linear pursuit. The genealogy underlying the evolution of meiotic sex takes the form of a multibranched bush, full of experiments. Some of the extant protoctists and other eukaryotes are clues as to how this bush might be reconstructed. In some groups meiotic sexuality, once acquired or evolved, became a prominent and irrevocable feature. In others it was secondarily lost, or was made extraneous by different processes that offered similar or greater utility. Finally, in some of the most ancient groups—those that branched off before meiotic sex had completely evolved—only pieces of the complex mechanism of meiotic sex are present.

MEIOSIS EVOLVES FROM MITOSIS

OUR PURSUIT of the origin of meiotic sex must begin with mitosis, the nearly universal mechanism by which eukaryotes replicate their DNA. Mitosis specifically involves the coiling of DNA with histone and nonhistone proteins to form chromosomes and the movement of these chromosomes on scaffolds of microtubules. The major exception to the universality of mitosis in eukaryotes may be Karyoblastea.

The only extant member of this lineage is *Pelomyxa palustris*—a giant, multinucleate amoeba, full of bacterial symbionts but lacking mitochondria. Although rRNA analysis has not yet been performed, *Pelomyxa* is a reasonable candidate for the title of direct descendent of the first eukaryote. Little is known of cell division in *Pelomyxa* except that its chromosomes do not appear to condense, and no spindles for mitosis have been identified. Mitosis has not, however, been entirely ruled out for Karyoblastea, as observations do indicate that the known extant member of this phylum has a brief stage with a nonmotile motility organelle.

Whatever the reproductive mode of *Pelomyxa* turns out to be, it is safe to say that mitosis must have evolved very close to the origins of eukaryotes. And as evidenced by its distribution throughout the extant eukaryote phyla, it must have been strongly selected for. The two major components of mitosis—the association of DNA with histone and nonhistone proteins to form chromosomes, and the movement of those chromosomes by way of microtubule spindles—must be accounted for in the early eukaryote. Also, the sequestering of DNA within a nuclear envelope, a defining characteristic of all eukaryotes, must have deep roots.

It happens that an extant archaebacterium, *Thermoplasma,* has some eukaryotic features that suggest a relatively direct connection with the early eukaryote host. One such feature is the presence of histonelike proteins associated with its DNA. *Thermoplasma* and another archaebacterial relative, *Sulfolobus,* also have flexible membranes. These two features, which may have been adaptations to the thermal, acidic springs in which *Thermoplasma* and *Sulfolobus* are found, were fortuitous preadaptations for a eukaryotic nucleus. That is, histonelike proteins, which seem to protect the DNA of *Thermoplasma* from hot acid hydrolysis, may have become the histones that help to bundle DNA into thick chromosomes. The membranes, especially flexible in *Sulfolobus* because of the hypothesized presence of steroidlike molecules that may themselves be protection against thermal stress, were preadaptations for the nuclear envelope.

Finally, the important tubulin protein must also have been acquired early in the history of the eukaryotes. A universal protein in eukaryotes, tubulin has been deployed for a number of different functions including spindle fibers. The acquisition of tubulin may have been via a symbiotic spirochete. (This controversial idea will be discussed in chapter 8.)

LONG STRANDS OF DNA NEED MITOSIS

MITOTIC CELL division seems to be a property of especially long strands of DNA, sometimes many orders of magnitude longer than prokaryotic DNA. The average prokaryotic genome is between 10^6 and 10^7 base pairs long. Protoctist and fungal genomes average between 10^7 and 10^8 base pairs, while the nuclear genomes of plants and animals fall between 10^7 and 10^{11} base pairs.[1]

[1] If a prokaryotic genome from one cell (10^6 base pairs) were stretched out, it would be about 1 millimeter long. If a eukaryotic genome (10^9) were stretched out, it would be about 1.5 meters long.

The prokaryotic genome seems to be particularly constrained in size and perhaps also in the arrangement of its genes. This is surprising, given the long and divergent history of prokaryotes, and the fact that they readily pick up extra DNA. In spite of an apparent limitation in genome size, prokaryotic genomes do show evidence of past duplication and acquisition events on the gene level (as evidenced by multiple copies of some genes) and on the entire genome level. *E. coli,* for example, has four nearly identical gene clusters at 12 o'clock, 3, 6, and 9 on its circular genome. The number and the symmetry of the repeated genes is taken as evidence of two genome duplication events (Riley and Anilionis 1978). Riley and Anilionis hypothesize that gene position may be conserved in prokaryotes because it serves some regulatory function. Prokaryotic operons (regulatory sequences) do, for example, often control several genes in tandem. This means that, in spite of a tendency to acquire DNA, there may be constraints on mixing, which would conserve not only genome order but genome size.

Furthermore, there may be some constraints on the size of prokaryotic genomes that pertain to the actual mechanism by which newly replicated DNA is partitioned to each of the daughter cells. The DNA attaches to a specialized protein on the cell membrane, and the two strands are separated from each other as that membrane grows during division. If indeed the four symmetrical gene clusters of *E. coli* are products of two duplication events, this suggests that there may be something intrinsically important about maintaining most of the cell DNA in one circular piece. In a sense, *E. coli* seems to have gone from haploid (single copy of the genome) to diploid (double copy) to tetraploid (four copies); but instead of four strands of DNA, there is just one.

Plasmids (extra circles of DNA acquired by various sexual mechanisms) are often lost in prokaryotes. That is, they may replicate but often do not become segregated equally into each of the daughter cells, perhaps because they do not attach reliably to the cell membrane, or because plasmid replication is not coordinated with the replication of the main genome. If a plasmid has no useful gene for the host, then it is a burdensome piece of DNA and it is to the advantage of the host to lose it. Thus plasmids may be present in multiple copies, having replicated on their own; or they may be lost if they fail to replicate when the cell divides (Lewin 1987).

An alternative to the idea that long strands of DNA need mitosis in order to replicate is the converse. Perhaps mitosis needs long strands in order to work efficiently. An intriguing piece of supportive evidence for this alternative explanation is that tiny, artificially produced chromosomes are much more easily lost when inserted into a mitotic system with normal-length chromosomes (Murray and Szostak 1985).

THE ROLE OF HISTONE PROTEINS IN MITOSIS

WE HYPOTHESIZE that the transition from small prokaryotic to large eukaryotic genomes may have involved histonelike proteins (which became eukaryotic histones) and tubulins, both of which may have loosened the constraints on size. It should be noted that while eukaryotic genomes are indeed large, the actual coding capacity remains small in that only about ten percent of any eukaryotic sequence consists of genes with a phenotypic expression. Therefore, the extravagant increase in genome size represents not so much an increase in actual gene number (although eukaryotes have at least one order of magnitude more genes than do prokaryotes). Rather, the larger genome reflects an increased tolerance for all kinds of extra DNA including repeat sequences and "nonsense" sequences.

This loosening of constraints may have eventually led to different regulatory mechanisms and even to the acquisition of some types of introns, but in the initial stages it may simply have been a tolerance of extra DNA. We hypothesize that this tolerance came about when DNA, already associated with histonelike proteins in the early eukaryote, came in contact with the newly acquired tubulin protein. Additional increments of DNA may have been not merely tolerable from the standpoint of structure but beneficial in some ways as well. One possible benefit pertains to the site of a replication origin (a place on the DNA strand where replication begins). As the genomes get larger, these origin sites must have duplicated many times. The resulting eukaryotes gained the ability to undertake many replication events simultaneously along each DNA molecule, thereby speeding up the process.

THE ROLE OF TUBULIN IN MITOSIS

TUBULIN HAS two convenient characteristics, its self-polymerization to form long tubules (microtubules) and its tendency (as an acidic protein) to stick to basic molecules. If tubulin genes were expressed in the early eukaryote, then two other evolutionary steps may have been nearly unavoidable—namely, the assembly of tubulin networks and their "decoration" with whatever basic proteins were available. These tubulin networks would have been some of the first components of the eukaryote cytoskeleton, and some networks might have become decorated with DNA via the basic histone proteins.

A rough visual analogy might be clotheslines (microtubules) strung between buildings from which laundry (DNA) dangles, attached by clothespins (histones). Histones hold the DNA in place and physically shorten it, an attribute that may have kept the growing genome a manageable size for cell division. The association of the genome with the tubulin cytoskeleton could have fortuitously guaranteed DNA segregation during cell division. This is because of an important property of cellular microtubules, and one that seems to support a hypothesis of symbiogenesis. Microtubules are initiated at microtubule organizing centers (MTOCs), which replicate in synchrony with the DNA of the host cell. (For details see chapter 8.) This semiautonomous replication of microtubules may be a remnant of the activity of the original symbiont, which would have experienced selection pressure to divide in synchrony with its host. Therefore, as the histone-bound chromosomes replicated in the early eukaryote, so would the MTOCs and in effect all of the cytoskeleton. The rigid conformation of the cytoskeleton might have helped guarantee a relatively even dispersal of the microtubules to the daughter cells, along with the attached chromosomes. A mechanism on the scale of (in effect) the entire cytoskeleton of the cell would have allowed an increase in genome size, while assuring a fairly even distribution of genes to daughter cells.

Refinements would have evolved later, including a molecular motor (or microtubule associated protein) hypothesized to facilitate the sliding of microtubules during chromosome movement. This would be analogous to the innovation of pulley systems for reeling in the laundry on clotheslines. The tubulin and other components of the cytoskeleton would also have allowed cells to grow larger, while still maintaining stability and shape. Thus one of the first indications of eukaryotes in the fossil record might be a marked change in cell size, although this remains a somewhat elusive piece of evidence in the sketchy fossil record.

THE ROLE OF THE NUCLEAR ENVELOPE IN MITOSIS

THE INVOLVEMENT of the nuclear envelope in mitosis is quite variable and suggests that the nuclear enclosure itself might be a later development than mitosis, and one that was integrated in a number of different ways among the protoctists (Margulis 1981). In general, mitosis may be either closed or open. That is, the nuclear membrane may remain intact during mitosis—as occurs in microsporans, zoomastigotes, euglenids, ciliates, oomycetes and others—or it may disassemble in the early stages of mitosis, allowing the chromosome movements to occur in the larger space of the cell. Nuclear membrane disassembly occurs in chrysophytes, cryptomonads, diatoms, and others. In some protoctistan phyla there are polar fenestrae, large openings in the nuclear envelope through which spindles pass; these groups include hyphochytrids and phaeophytes. Many phyla show a variety of these mitotic styles (Margulis 1981; Raikov 1982).

It is interesting that the two phyla which appear to be early branches of the eukaryotes, the microsporans and the diplomonads (of the zoomastigotes), both have intact nuclei during mitosis. It may be, however, that there was no strong selection pressure in most groups for a particular treatment of the nucleus in mitosis, given the diversity displayed.[2]

MEIOTIC SEXUALITY AND THE ROLE OF ENVIRONMENTAL STRESS

MEIOTIC SEXUALITY involves two components: the fusion of two cells, bringing two haploid genomes in contact (thereby forming a double or diploid genome in a single cell), and the subsequent division, meiosis, by which the doubled genome is reduced to its original haploid state. Meiosis is utilized by many protoctistan phyla, and it occurs in most (but not all) plants, fungi, and animals. The many

[2]For example, dinoflagellates, which have been considered primitive in respect to their style of mitosis, now appear to have just an unusual variation on mitosis, as their rRNA places them on a recent branch close to ciliates on the family tree.

protoctistan phyla either lacking meiotic sexuality or undergoing meiotic sex only rarely during cell reproduction are testaments to the nonessential or optional nature of sexuality. The exceptions, those organisms for which meiotic sex is obligate (being linked to reproduction) are truly exceptional, albeit highly visible, as they include among other groups the vertebrate animals.

The sequence of events by which meiotic sex evolved has been proposed by Margulis and Sagan (1986). As with bacterial sex, environmental stress may have been the major selection pressure for the first step in meiotic sex, cell fusion. Organisms deal with environmental stress in many ways, including movement away from stress, formation of resistant protective structures, and cell fusion. Cell fusion, itself, comprises a variety of mechanisms including (if food is scarce) cannibalism.

Many protoctistan phyla use one or more fusion strategies when stressed—with or without any accompanying meiotic sex. When the ciliates *Blepharisma* and *Tetrahymena* are starved, individuals begin to feed upon their neighbors, becoming giant in size. The zygotes of *Dictyostelium,* a cellular slime mold, are formed by the fusion of two haploid amoebae. These zygotes go through a cannibalistic phase, perhaps a remnant of an earlier form of sexuality. The zygote attracts other amoebae to it and together they form a protective cyst. Inside the cyst, the zygote eats the other amoebae and becomes a "giant cell." Just before the cyst breaks apart, the giant cell divides, so that several amoebae emerge.

Cannibalism would be a logical mechanism by which two cells of the same type might get their nuclei in close enough proximity for fusion. Organisms have numerous mechanisms for recognizing their own DNA and protecting it from digestion. For example, DNA sequences may have methyl groups attached in specific places to prevent digestion by host endonucleases. Such specific mechanisms may have protected some or all of a cannibalized organisms's DNA long enough for it to recombine with the host genome. In fact, some actual fertilization events look rather like cannibalism. When hypermastigotes (zoomastigotes from termite and wood roach intestines) mate, one individual is taken up entirely by the amoeboid posterior end of the other. The hypermastigote inside appears to be digested, leaving just the chromosomes that become part of the new diploid genome (Cleveland 1947). When two haploid cells of *Chlamydomonas* (a chlorophyte) fuse to form a gamete, usually one of the chloroplasts (a single, large horseshoe-shaped structure) is digested, reminiscent perhaps of a cannibalistic origin for this fusion.

THE EVOLUTION OF GAMETES

CANNIBALISM, LEADING to opportunities for DNA recombination, might have become fixed fairly quickly in populations—especially if a new combination of genes conferred a selection advantage under stressful conditions to the cannibal "partners." One particular stress might have been that of acquiring symbionts, some of which might have been quite pathogenic. The apparent ease with which the eukaryotic host acquired mitochondria and plastids might have had a downside. The innate hospitality of the host might have made it easy for unhelpful or harmful symbionts to take up residence. Sexuality (resulting in new gene combinations) might have helped to make the cytoplasm of an individual convert quickly to inhospitality to a pathogenic symbiont (Hamilton et al. 1990).

In those species for which DNA recombination attendant to fusion became as important as the nutritive value of the partner, there may have been selective pressure for one of the cells in the partnership to become smaller. Some groups of protoctists, as well as plants and animals, exhibit striking differences in the size of their gametes (haploid cells). The microgametes have traditionally been defined as male (sperm, pollen) and the macrogametes are defined as female (eggs, ova). More precisely, the male gamete is defined as the donor, the female as the receiver. Many protoctists and most of the fungi, however, maintain gametes of equal size (isogametes), with no particular designation of maleness or femaleness. Selection pressures for microgametes may have been for ease of motility and dispersal, as microgametes are often motile or light enough to be carried by wind, for example, as in plant pollen. Selection pressures for macrogametes would have pulled in the opposite direction, as stripped-down sperm offered less and less nutrient resources along with the gift of DNA.

As newly sexual organisms became more and more receptive to DNA recombination brought by some form of fusion, there may have been selection pressures for restricting fusions to cells that were similar enough to have the same set of genes (assuring even recombinations) but not so similar that the genes were identical in every respect (i.e., clones). The advantage of recombination in a stressful environment would be the possibility of bringing together alleles that individually would not be advantageous but together would be so. The need to find a complement would mean that a partner ought to have the same functional locus, but it would also mean that it would be advantageous to screen out partners that are too self-similar. Thus may have evolved *mating types,* defined simply as cells that are compatible for mating with other related cells but not with cells of the exact same type. The concept of mating types is somewhat different from that of maleness and femaleness, which has more to do with motility and the donor-recipient distinction. Mating types, in addition, can number more than two.

Mating types seem to prevent inbreeding—the deleterious accumulation of identical genes, some of which, if found in more than one copy, may be lethal. In some cloned, experimental rat colonies, virtually identical rats reproduce compatibly generation after generation with typical mammalian sex. These lab populations are extremely inbred and, by wild standards, unfit. In most cases they must be maintained with special care.

Paramecia are good examples of organisms with multiple mating types. Two paramecia may or may not attach together to exchange nuclei, mouth part to mouth part, depending on which of eight or more mating types they are. Fungi may have hundreds of mating types; *Schizophyllum* has about 340 (Kendrick 1985).

It may be then that the evolution of motile microgametes and relatively nonmotile macrogametes (that is, maleness and femaleness) took a different pathway, and because of decidedly different selection pressures, than did the evolution of isogametes with specific compatibility requirements. In the first case the emphasis seems to be on dispersal via motility, such that by virtue of distance from the parent an individual microgamete might be nearly guaranteed an unlike partner. In the case of isogametes, in which both partners are equally motile or nonmotile, recognition of preferred mating types is important. Of course, there are examples of organisms in which both strategies are at work.

Many plants, for example, have mechanisms for incompatibility with themselves, so that pollen raining down onto the stigma from the same flowers do not self-fertilize. These plants also have mechanisms for fostering dispersal by, for example, attracting insect pollinators. There are many cases, too, in which self-fertilization or fertilization with identical clones does not seem to be prevented by any compatibility mechanisms, as in the case of inbreeding cloned laboratory rats and many plants, such as some fruit trees. These departures from the "rule" are, in part, what makes it difficult for researchers to demonstrate conclusively that meiotic sexuality is particularly advantageous. This has to be analyzed on a case by case basis and for specific, often stressful environments.

MEIOTIC DIVISION: WHY IS IT NEEDED?

MEIOTIC CELL division is the mechanism by which a cell that has doubled its genome by fusing can then reduce its genome to the original size. Since it is this pattern, fusion followed by meiosis, that we see again and again in the sexual cycles of most eukaryotes, there must have been some advantage for the ancestral meiotic organisms. To state it simply, a series of cell fusions not followed with meiosis would quickly enlarge a genome with dozens and then hundreds of extra copies of each chromosome.

The advantages of being diploid (the possibilities of complementation, protection from mutation) might, if accelerated to polyploidy, be a disadvantage in reproduction. DNA polymerase replicates a eukaryotic chromosome at approximately 2000 bases per minute.[3] A twofold increase in number of chromosomes would slow cell division, but apparently in many organisms this disadvantage in speed is outweighed by diploid advantages. A one hundred–fold increase, however, might significantly slow reproduction with detrimental effects. A one hundred–fold increase might also significantly crowd the division apparatus (spindles and MTOCs) to reduce efficiency.

CLUES DRAWN FROM THE EXCEPTIONAL PLANTS

IT IS somewhat risky to generalize about the importance of meiotic division, as many organisms are polyploid and the plant kingdom in particular has a remarkable rate of polyploidy. In plants, polyploidy seems to be due in part to the relatively loose specificity for mating.

[3] *Drosophila* 2,600 bp/min; *Xenopus* 2,200 bp/min (Lewin 1987).

Pollen from a close but different plant species can sometimes fertilize an ovum and produce a zygote that cannot undergo meiotic division because the sexual partners had different numbers of chromosomes. This seems to do many plants no harm and may even confer an advantage.

Indeed, many hybrid crop plants are bred to be polyploids, with no expectation that they might be able to undergo normal meiosis. This is why viable seeds usually cannot be obtained from hybrid garden crops. The advantage of polyploidy to these particular plants is sometimes accompanied by a distinctive vulnerability: they are utterly dependent on a particular species, humans, for their propagation. Polyploidy in crop plants has been observed to increase size of various plant structures, a benefit that seems to outweigh slower growth and reduced fertilization (Stebbins 1971).

Acquisition of extra chromosomes that then become a part of the regular diploid number of chromosomes also has been an important feature in plant evolution. Stebbins (1971) suggests that all plant genera with diploid numbers greater than twelve are probably products of polyploid events, in which the additional chromosomes eventually became part of the diploid number. An extraordinary example of at least six chromosome doublings is that of the fern *Ophioglossum reticulatum*. With about 630 chromosomes, *Ophioglossum* has a crowded nucleus and may represent an upper limit of the capacity of the mitotic apparatus to handle large numbers (Stebbins 1971). Thus, the following discussion on the importance of maintaining certain haploid and diploid states in a sexual cycle should be taken with the understanding that it may not be true for all sexual organisms.

The development and then persistence of a meiotic cycle in the first sexual eukaryotes may well represent a crucial selection pressure, perhaps for something as simple as swiftness of DNA replication. In many of the extant sexual groups, however, sex may be a vestige preserved for altogether different reasons. In some plants such as dandelions, which are triploid, the superficial aspects of sexuality are maintained; dandelions produce flowers and seeds. But the flowers are for naught, and the seeds are essentially produced asexually in this successful, cosmopolitan plant. Why then does the dandelion waste energy in producing a flower and even going through the motions of meiosis? The answer resides in the complexity of most of the plant's organismal traits (semes). The development and dispersal of dandelion seeds has long been associated and integrated with a sexual cycle that was once functional. Since part of the whole integrated system is essential, namely the mobile seeds, the rest of the system is maintained.

A similar example is that of most mammals, which lack any asexual alternatives for reproduction, such as division, budding, and asexual spore formation. Meiosis in this case has become intimately involved with the only replication and dispersal system available, and thus has been irrevocably preserved in the mammalian line. In both cases, dandelions and humans, sexual traits have been fixed in the lineage for entirely different reasons, and both reasons surely differ from the selection pressures that account for the origins of eukaryotic sex.

Stephen Jay Gould (1991:66–75) has a fine analogy in this regard. The QWERTY keyboard originated in manual typewriters. It was deliberately designed to be awkward enough to prevent typists from speeding and thereby jamming the hammers that embedded their symbols through the inked ribbon. Now, although it is obviously impossible to jam the lettering devices of word processors, the QWERTY system persists; why? For no better reason than that a change in system would render millions of typists and keyboards alike instantly obsolete. Thus (although it is possible to convert a keyboard to a more modern configuration) most of us retain the QWERTY system. Likewise, in some organisms sexual traits are as embedded as the QWERTY system, and extinction (lack or reproduction) would be the consequence of eliminating sex.

MEIOSIS AS A MODIFICATION OF MITOSIS

MEIOSIS, THEN, as the mechanism by which haploid number is restored after fertilization, might have evolved in those early eukaryotes for which speedier division conferred a real selection advantage and for which accumulating copies of the genome would be a disadvantage. At least three major steps must have occurred by which the mitotic (cell division) machinery of eukaryotic cells was converted for use in meiotic division.

The simplified steps of mitotic cell division are as follows (see also figure 5.1):

1. Replication of the MTOCs (microtubule organizing centers) that will eventually act as organizers for the spindle fibers.

2. Replication of DNA such that the old and new strands remain attached to each other by a shared centromere.
3. Splitting of the centromere so that the old and new DNA strands may move to opposite sides of the cell by the spindles (having been organized by the MTOCs).
4. Division of the cell into two cells.

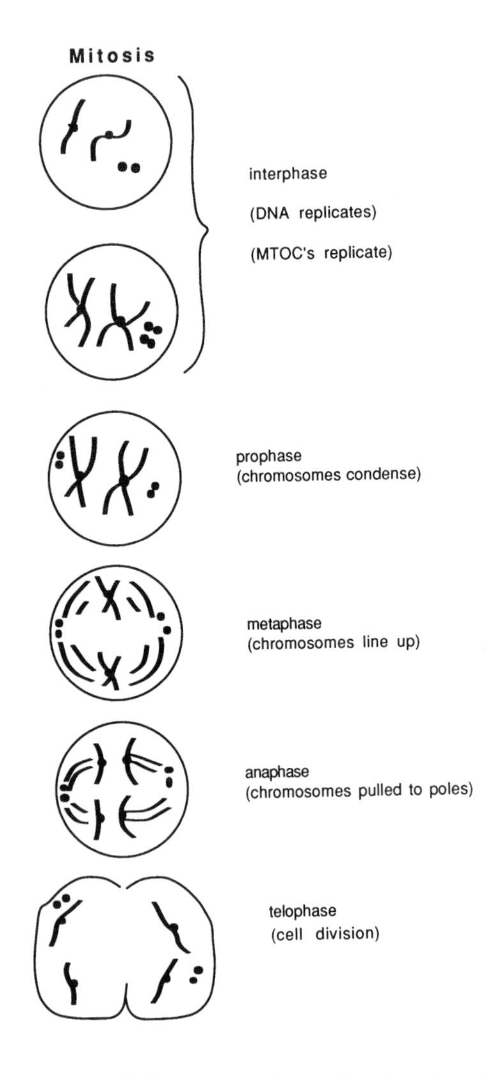

FIGURE 5.1. Mitotic cell division is the mechanism by which eukaryotic cells reproduce.

Meiosis is a modification of three of those steps. First, the MTOCs divide (precociously) twice during the cycle, but there is no second duplication of DNA. Second, synapses (special connections) are formed between like chromosomes, causing the chromosomes to line up differently. Finally, the splitting of the centromere is delayed until the end of the entire meiotic cycle.

Thus, the simplified steps of meiosis can be depicted as follows, with the crucial variations marked by italics (see also figure 5.2):

1. Duplication of the MTOCs that will eventually act as organizers for the first spindle.
2. Replication of DNA such that the old and new DNA strands remain attached to each other by a shared centromere.
3. *Formation of synapses between like chromosomes, which seems to enhance the ability of the chromosomes to recombine with each other.*
4. *Disintegration of the synapse, at which point like chromosomes move to opposite sides of the cell by the spindles. Note: centromere splitting has been delayed.*
5. *Division of the cell into two cells.*
6. *A second replication of the MTOCs, but no DNA replication.*
7. Splitting of the centromeres so that old and new DNA strands may move to opposite sides of the cell by the spindles.
8. Division of the cell into two cells.

The specific modifications that led to steps 3, 4, and 6 may be observed in organisms (yeast and fruit flies) carrying mutations that cause one or more of these modifications (cell division mutations). These key modifications are also reflected in the normal division modes of certain protoctists that retain vestiges of modified mitosis and meiosis in their cell division.

CELL DIVISION MUTATIONS IN YEAST

YEAST ARE among the more useful organisms for the study of cell division cycle mutations. Under most circumstances, mutations causing a failure in a step of cell division would be lethal in an organism. However, in yeast some of these mutations are temperature sensitive so that the lethal effects are observed only at high temperatures. Thus strains of cell-division mutants can be main-

FIGURE 5.2. Meiotic cell division occurs only during eukaryotic sex for the purpose of reducing the number of chromosomes by half.

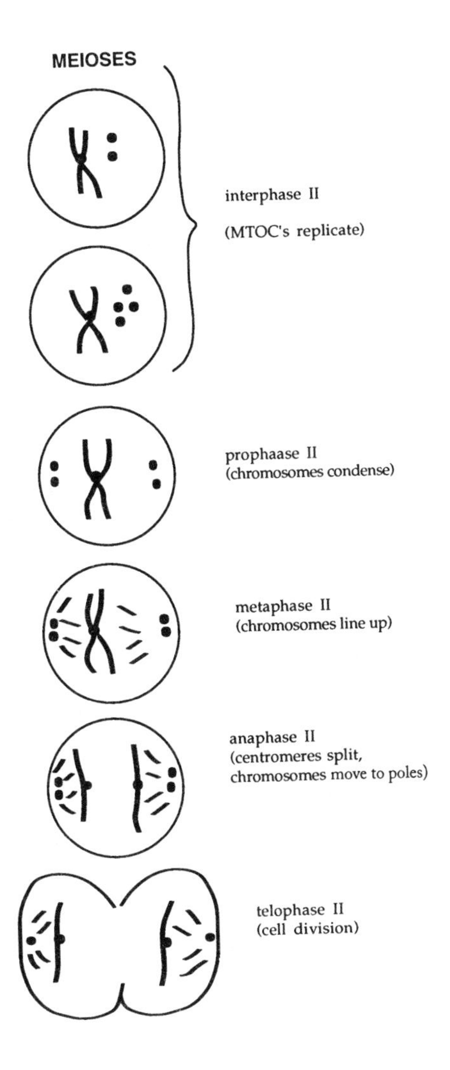
MEIOSES
interphase II
(MTOC's replicate)
prophaase II
(chromosomes condense)
metaphase II
(chromosomes line up)
anaphase II
(centromeres split,
chromosomes move to poles)
telophase II
(cell division)

tained at sufficiently low temperatures for cell divisions to occur (Hartwell 1978).

The mutants may be divided into seven classes, some of which may be especially relevant to the conversion from mitosis and meiosis:

1. duplication of MTOCs
2. separation of MTOCs
3. initiation of DNA Synthesis
4. DNA Synthesis
5. separation of the chromosomes (centromere splits)
6. division of the cytoplasm
7. final separation of the cells

In particular, a mutation such as spo 13 that slows the rate at which MTOCs are duplicated has the effect of omitting the final meiotic division, leaving the mutant yeast with a diploid cell. It is hypothesized that in yeast mitosis, extra cell divisions are suppressed and that in normal meiosis this suppression is somehow omitted. The spo 13 meiotic mutants seem to be experiencing the mitotic-like suppression (Murray and Szostak 1985). Although many such mutations would be lethal to the organism, some combination of mutations might have been sufficient for a conversion of mitosis to a meiotic-like division in ancestors of yeast. Because it is primarily timing of certain steps that makes the difference between mitosis and meiosis, the conversion could occur via mutations that prompt the usual mitotic sequence of events to occur in a slightly different order, owing perhaps to a different order in the sequence of gene expression (Hartwell 1976).

SYNAPSE MUTATIONS IN FRUIT FLIES

THE FORMATION of synaptonemal complexes, which briefly hold together similar (homologous) chromosomes during the early stages of meiosis, is unique to meiosis,[4] but it may have its ancestral roots

[4]Or it is nearly unique. Stack and Brown (1969) review numerous observations of structures resembling synaptonemal complexes, which hold together homologous chromosomes at various other stages in the cell cycle, including mitosis. Whether or not these structures are the same as meiotic synaptonemal complexes remains to be determined.

in one of the many mechanisms by which DNA is repaired. One of the basic repair mechanisms in all cells, and which might have been an important preadaptation in bacterial as well as eukaryotic sex, occurs when similar chromosomes exchange sequences or when a missing sequence is patched in from one that is complete (chapter 3). The synaptonemal complex enables numerous recombination events to occur between chromosomes by holding them in close proximity and correct alignments.

Large protein complexes called recombinational nodules seem to mediate the actual events of recombination. Mutant female fruit flies (*Drosophila*)[5] with fewer than usual or irregularly spaced nodules have correspondingly fewer or more irregular recombination events (Carpenter 1975). Some of the synapse mutants of fruit flies produce gametes with irregular numbers of chromosomes (Murray and Szostak 1985). Synaptonemal complexes also hold the chromosomes together in a position such that like pairs may be easily separated from each other during the first division. This, however, may be a secondary and somewhat serendipitous function for these complexes in which recombination was the primary function. Synaptonemal complexes may even have been a somewhat later development in meiosis, as evidenced by some of the strange meiotic variations that will be described in the next section.

EXTANT VARIATIONS ON MEIOSIS

THE PROTOCTIST phylum Zoomastigina is one of the more interesting places to study the origins of meiotic sexuality. This phylum includes the diplomonad and retortamonad classes, both of which lack meiotic sex (Vickerman 1989; Brugerolle and Mignot 1989). On the basis of their lack of mitochondria and in the case of the diplomonads, the 16S RNA analysis,[6] these two classes may be descendants of the earliest eukaryotes.

Two other classes of Zoomastigina that lack mitochondria, the

[5] Curiously, the reproductive cells in male fruit flies do not form synapses and do not recombine their chromosomes. They still manage, however, to accomplish meiotic divisions. In bumble bees, it is the female reproductive cells that fail to form synapses.

[6] Some of the microsporans do some type of meiotic sex; this group is also considered to be directly descended from early eukaryotes on the basis of 16S RNA.

parabasalids and pyrsonymphids, have members that undergo what may be very early versions of meiosis. L. R. Cleveland (1947) worked out a plausible sequence for the origin of meiosis based entirely on variations that he observed in these two classes. An initial step may be observed in genera of Parabasalids, such as *Holomastigoides* which seems to be a permanent diploid. *Holomastigoides* presumably doubled its chromosomes at one point, but lacked a meiotic step to reduce them. There is evidence in other parabasalid genera that polyploidy has occurred and that this has resulted in both larger diploid numbers and in aneuploidy (irregular numbers of chromosomes).

For example, if *Holomastigoides* has two sets of two chromosomes, presumably an ancestral number, then species with two sets of three chromosomes or two sets of five chromosomes and even three sets of five chromosomes may represent polyploidy events. A particularly convincing sequence of polyploid events begins with a species of *Spirotrichosoma* with 12 chromosomes, followed by other species in this genus with 24, 48, and 60 chromosomes. This is reminiscent of the sequences of polyploid events that can be observed in plants. These various parabasalids, then, represent descendants of several early experiments in chromosome doubling with no subsequent meiotic reductions. Cleveland noted that greater numbers of chromosomes seemed to represent greater difficulties in reproduction. Cell division in these organisms looks like the first division of meiosis and even includes what appear to be synaptonemal complexes, except that the MTOCs and the chromosomes duplicate in mitotic synchrony. The result is no change in ploidy.

Another variation is represented by *Barbulanympha*, which duplicates its chromosomes from haploid to diploid and then undergoes a one-step meiosis to return to haploidy. In a sense, *Barbulanympha* has a solitary sexual cycle (autogamy) in which it goes from haploid to diploid to haploid without any participation of a sexual partner. The initial duplication of chromosome occurs without a nuclear division because the centrioles in this case are enormous and take two days to mature after their replication. During this time the chromosomes have an opportunity to divide yet again, yielding a set of four. Finally the mature centrioles perform their task of separating the quadrupled chromosomes into two diploid cells. The next division is the meiotic one, but this time the centrioles have a head start and are mature before the chromosomes are ready for them. Thus the final cell division reduces the diploid number to haploid.

The pyrsonymphids and some of the parabasalids also do a one-step version of meiosis but have a sexual partner. For example, *Oxymonas* is a haploid pyrsonymphid. After fertilization (the fusion of two haploid cells) the diploid cell commences meiosis without a DNA replication step. Thus the single division is sufficient to reduce the diploid to two haploid cells.

A nontraditional two-step meiosis (actually, traditional compared to the odd variation of *Barbulanympha*) is performed by other parabasalids, such as *Holomastigoides.* However, *Holomastigoides* is also capable of undertaking a type of mitosis that includes some meiotic features, such as pairing of like chromosomes. Also, *Holomastigoides* seems to spend a large part of its division cycle with four of each chromosome.

Although the evidence tempted L. R. Cleveland (1947) to draw such a conclusion, these variations on meiosis (which are just a few of the possibilities; see Raikov 1982) should not be considered as steps on a linear pathway to so-called normal meiosis. Rather, the diversity of meiotic expression is indicative of the degree of tolerance organisms may have for changes in their pattern of mitotic and meiotic divisions. Two-step meiosis, as it is traditionally diagrammed for animals, plants, fungi, and a few protists, must have been one of those many variations—but a particularly successful one or at least one that became quickly fixed in the population. The Zoomastigina are probably not unique except that their chromosomes are large and their divisions are especially easy to study. If other protists yielded up the secret details of their divisions as easily, there would be an even greater list of variants known.

SUMMARY

THE EVOLUTION of eukaryotic sex is often described mainly or even entirely with examples from the animals, many of which have their reproductive cycles obligately linked with sexuality. This chapter has taken most examples from the protoctists, as well as plants and fungi, for the purpose of showing how the early evolution of sex might have occurred. In these less-studied kingdoms, reproduction without sex seems to be far more common. Nevertheless, sexuality seems to have evolved early and to have branched off in a diversity of ways in many groups.

THE EVOLUTION OF EUKARYOTIC (MEIOTIC) SEX

This chapter has also described the evolution of mitotic division, with histones and tubulin as essential components. Sexual cell division (meiosis) seems to have evolved from mitosis and the events may be dissected with the help of meiotic mutants (especially in those organisms like yeast for which sex is not obligate). How and why cells evolved and maintained the mechanisms by which they fuse together sexually is not well understood. Some extant protoctist groups provide clues in their unusual variations on fusions. This diversity, especially in protoctists, may provide some of the most intriguing evidence for the early evolution of sex.

VI

The Acquisition of Mitochondria

MOST EUKARYOTES have mitochondria, specialized compartments in which respiration occurs. Mitochondria were acquired as symbiotic bacteria by eukaryotic (nucleated) cells probably about 2.5 billion years ago. Evidence for this symbiotic origin includes data and observations from several fields of biology.

Throughout this chapter (and the next two chapters on the acquisition of organelles for photosynthesis and motility), we will examine evidence that three organelles in eukaryotic cells arose from previously free-living prokaryotes that entered into symbiotic associations with the larger host cell. Specifically, symbiogenesis as a theory of organelle origin gains support if:

1. enzymes and protein complexes within the organelle are more similar to prokaryotic than to eukaryotic cytoplasmic analogues.
2. a free-living prokaryote can be found that has strong genetic, biochemical, and morphological resemblances to the organelle.

3. the organelle itself retains a genome, with characteristics more similar to those of a prokaryotic than a eukaryotic genome.
4. ribosomal RNA, transfer RNA, and messenger RNA within the organelle are more similar to those of prokaryotic than of eukaryotic cytoplasm.
5. the organelle has an ability to replicate and a genetics that is separate from the nuclear genetics.
6. the organelle is found in eukaryotes as an "all or nothing" phenomenon in which one either finds the organelle as a whole (or in some modified form) or does not find it at all (if it has not been acquired or was secondarily lost); one does not expect to find intermediate stages of the organelle if it was acquired all at once as a symbiont (see also appendix B).

OXYGEN NOT A PRIMARY SELECTION PRESSURE

EARLY FUNCTIONAL explanations for the symbiotic origin of mitochondria argued that increasing oxygen in the atmosphere selected for certain anaerobic host bacteria (or early eukaryotes) to acquire respiring bacteria as symbionts. Such a relationship would have the advantage of detoxifying oxygen for the host while introducing a new and efficient form of metabolism, respiration (Sagan 1967). This was a useful working hypothesis for many years; today, however, it is has been modified to take account of what is now understood of the metabolism of aerobic relatives, *Thermoplasma* and *Sulfolobus.* Also, the eukaryotic cytoplasm itself is not completely anaerobic but has various oxygen detoxification mechanisms, which were probably already in place at the time of the symbiosis.

Earth's early atmosphere was anoxic and all organisms were anaerobic. As the molecular or "free" oxygen in the atmosphere gradually increased to the current level of twenty percent, anaerobic prokaryotes evolved mechanisms to cope with oxygen. Oxygen is a corrosive molecule which readily oxidizes organic compounds and also forms reactive and therefore toxic superoxides (O_2^-), hydroxyls (OH^-), and peroxide (H_2O_2). Diverse mechanisms for coping with these compounds evolved in prokaryotes; they include enzymes such as peroxidases, catalases, and superoxide dismutases that disarm the toxic molecules by using them in controlled oxidation reactions. Light exacerbates the toxic effects of oxygen in a process called photooxidation. Isoprenoids and their derivatives (such as the many pigmented forms of carotene) protect against photooxidation by blocking light. Also, as some of the isoprenoid compounds require oxygen in their synthesis, this constructively detoxifies oxygen by using it for a biosynthetic purpose. Synthesis of steroids (from the isoprenoid pathway) and of polyunsaturated fatty acids are processes limited to aerobes (most eukaryotes and a few prokaryotes) and may therefore be related to the availability of molecular oxygen (Margulis 1981). Perhaps the ultimate detoxification mechanism was so efficient that oxygen became a necessity for some organisms. For many respiring organisms today, oxygen serves as a terminal electron acceptor; this form of metabolism is characteristic not only of most eukaryotes and their mitochondria, but of some prokaryotes.

Very few prokaryotes remain without detoxification mechanisms for oxygen, and these are thus tied to anaerobic environments such as deep sediments. *Thermoplasma* and *Sulfolobus* both fall into the category of prokaryotes that have developed detoxification mechanisms. In fact, those of *Thermoplasma* are remarkably similar to the ones found in the cytoplasm of eukaryotes; that is, the type of primitive respiration that occurs in eukaryotic microbodies (peroxisomes) is present in *Thermoplasma* (Searcy and Whatley 1984). Microbodies, which are membrane-bound organelles, are characterized by the presence of oxidases that reduce molecular oxygen to peroxide in the course of oxidizing a particular substrate, such as formate. Then catalase reduces peroxide to water. Thus, one of the functions of this pathway in eukaryotes, and perhaps in *Thermoplasma,* is the detoxification of oxygen as well as the breakdown of specific organic compounds.

Overall, then, because early eukaryotes probably already possessed mechanisms by which molecular oxygen could be stripped of its toxicity, it is unlikely that oxygen detoxification was a primary motivator for an early eukaryote to form a symbiosis with the respiring, free-living ancestor of mitochondria. What, then, was the motivator?

ACIDIC WASTE PROBLEMS

SEARCY AND Whatley (1982, 1984) propose that the early eukaryotic cell, as with *Thermoplasma* today, had a problem with acidic waste products. Although *Thermoplasma* has some of the Krebs cycle enzymes, it apparently does not use any Krebs cycle products in an electron transport system to produce ATP. Its form of metabolism is fermentative and produces acetic acid as a waste product. *Thermoplasma* thrives in a pH of 1–2, partly generated by its own waste products, because it has a mechanism for keeping excess H^+ ions (protons) out of its cell. The mechanism is similar to the electron transport chain of respiration but apparently uses only cytochrome b and quinone. Cytochrome c, which is found in mitochondria and in certain respiring bacteria, is not found in *Thermoplasma*. The function of this respiration chain seems to be the export of H^+ ions. ATPs are not produced as they are in the more elaborate respiratory chain of respiring bacteria and mitochondria. *Thermoplasma* is therefore an obligate aerobe, in spite of using fermentative pathways for its metabolism, because oxygen is required for its H^+ ion exporting mechanism.

What is likely, then, for an early eukaryote is a selection for symbiotic associations that helped it deal more efficiently with its own acidic waste products. Such an association might have occurred casually at first, as a consortium of the eukaryote with a respiring bacterium. Searcy and Whatley (1984) note that when *Thermoplasma* is grown with a respiring "contaminant" the cultures grow much better, presumably because the partner removes some of the acetic acid. In fact, acetate would be treated as a food molecule by most respiring bacteria and so there might well be a strong selective advantage for partnership in dealing with the "problem" of acetic acid.

At least one bacterial consortium is known to involve a waste product problem. *Methanobacillus omelianskii*, though the name implies a single species, is actually a consortium of two bacteria—one of which produces hydrogen as a waste product while the other oxidizes hydrogen (Schlegel and Jannsch 1981). It makes sense that bacteria in their natural habitats would tend to cluster together if some were producing waste products that others could use. Although many such associations have been described (Schlegel and Jannsch 1981), many others have probably been ignored—owing to the passion of Western microbiologists for pure cultures.

If a consortium between a respiring bacterium and a eukaryote ancestor was the first step in the acquisition of mitochondria, then the second step must have been greater intimacy by the internalization of the respiring partner. The acetic acid produced by the glycolysis (fermentation) of the eukaryote host could then have been channeled directly to the premitochondrial symbiont. Releasing ATP from the symbiont to the host would have been a subsequent step. It is a step that might have arisen by chance but which would enhance the partnership by making the host that much more efficient in its environment.

INTERNALIZING THE SYMBIONT

HOW THE internalization step might have occurred is the topic of at least two hypotheses (Margulis 1981). One hypothesis is that the respiring partner was gradually taken in via the invagination of the host cell's wall-less membranes. Such invagination would not have to be an active, eating function such as what one sees in feeding amoebae. If the two bacteria were attached (as many consortia are) then a gradual shift of the attachment membrane from outside to inside would have been sufficient. For example, *Sulfolobus* is a rather irregularly lobed cell with a potential to form internal membranes from invaginations of its outer membranes. Nevertheless, active amoeboid activity should not be completely ruled out, as it appears that the cytoskeletal components necessary for endo- and exocytosis may already have been in place at the time that mitochondria were acquired.

A second hypothesis is that the respiring ancestor was one of a type of bacterium with a predatory habit, capable of entering another bacterial cell and digesting it from the inside out. Some extant examples, such as *Bdellovibrio,* actually penetrate only into the periplasmic space (between wall and membrane); others, such as *Daptobacter,* enter the cytoplasm (Guerrero et al. 1987). The fact that the best studied of these bacteria are quite destructive to their hosts does not detract from the possibilities of this being a mechanism for introducing an internal symbiont. There is considerable variability in respect to the speed with which *Bdellovibrio* reproduces and destroys its host and variability in the number of offspring a single invader can produce (depending upon the host). The extent of this range is unknown because researchers are for the most part hampered by their experimental methods. That is, predatory bacteria are most easily detected when they are killing other bacteria, sometimes even producing viruslike plaques on plates. Predatory bacteria that kill very slowly or simply enter the host and divide in approximate synchrony with it may well exist. They would represent an extreme in the range of predatory bacteria and in the range of host resistance mechanisms.

Whether or not the bacterial ancestor of mitochondria entered on its own power or was taken up, its establishment inside the host was a major step and one that must have been followed by numerous refining steps—steps by which the symbiotic partners could better control each other and maintain the relationship.

THE GENETICS OF MITOCHONDRIA

THE UNIQUE genetic system that mitochondria carry is one of the most compelling pieces of evidence for their symbiotic origin. The mitochondrial genome consists of one or more circular pieces of DNA or, in ciliates, linear DNA (for a review, see Prebble 1981). The length of the mitochondrial DNA ranges from 500,000 base pairs in plants to about 20,000 base pairs in animals. This represents a considerable loss of DNA compared to the bacterial ancestor of mitochondria, which we presume had 1–2 million base pairs of DNA, typical of most prokaryotes. Another piece of evidence in support of symbiogenesis is that mitochondria contain a distinctive DNA polymerase—different from that in the nucleus—which is used to replicate mitochondrial DNA. Mitochondria also use a distinctive RNA polymerase to transcribe mRNA from the DNA.

The ribosomes of mitochondria are generally different from the ribosomes of the cytoplasm. For example, mammalian ribosomes have sedimentation coefficients (S: estimates of molecular size based on density) of 80 S, while the ribosomes of mammalian mitochondria are 55 S. Furthermore, mitochondrial ribosomes differ in size ranging from 55 S in mammals to 80 S in *Tetrahymena*, a ciliate. The cytoplasmic ribosomes are relatively constant in size, 77–80 S. Translation occurs on the mitochondrial ribosomes using a version of the genetic code that differs markedly for some codons from the so-called "universal" genetic code. (The evolution and significance of this alternate code will be discussed in the next section.)

Mitochondria use their genetic systems semiautonomously to code for and produce protein products and to replicate their genome during division. New mitochondria are produced when preexisting mitochondria replicate their DNA, divide, and grow. This is in contrast to nonsymbiotic cell structures such as lysosomes, which are formed of invaginations of the endoplasmic reticulum. In the case of yeasts, which have a single large mitochondrion, this replication takes the form of budding and branching rather than the many individual mitochondrial divisions that occur in most cells. The number of mitochondria per cell is maintained fairly predictably owing to the synchronous divisions of mitochondria at a time when the host cell divides. However, mitochondria may be dividing, fusing, and budding at other times in the cell cycle—although still maintaining a stable number of mitochondria in the cell.

SHARED CODING

MITOCHONDRIA ARE not genetically autonomous because many of the proteins they require are encoded by the nucleus and imported into the mitochondria. In some cases, as for large enzyme complexes, some subunits are coded for by the mitochondria and some by the nucleus. This shared genetics is one of the most intriguing pieces of evidence for a symbiotic origin. It suggests that at some point in the history of the association, genes were transferred horizontally from mitochondria to nucleus and probably from nucleus to mitochondria—and, if plastids were present, mitochondria to plastid and vice versa. Although there are mechanisms by which genes may be moved, such as in viruses, it is not usually possible to identify the mechanisms by which specific mitochondrial genes moved.

The shared coding for large multiprotein complexes, such as the F_1 ATP synthetase, remains the best evidence that gene movement occurred. In this case the mitochondria probably entered the symbiotic association with the entire coding capacity for the F_1 ATP synthetase and subsequently lost some of the genes to the nucleus. It is interesting to note that the specific genes transferred differ in the cases of two fungi, *Saccharomyces* and *Neurospora.* In the former the mitochondrion codes for subunits 6, 8, and 9 of the ATP synthetase and in the latter, only subunits 6 and 8 (van den Boogaart et al. 1982). *Neurospora* does, however, carry a pseudogene (nonfunctional gene) for subunit 9. This is evidence for how part of the transfer sequence might have occurred.

First, a copy of the gene for subunit 9 was made by an unknown duplication mechanism. A virus or viruslike piece of DNA might have accomplished this task, or the duplication could have resulted from an uneven DNA recombinational crossover event. Unfortunately there are no signatures left for this event; genes transferred from yeast mitochondria to nuclei are near transposons (jumping genes) called "Ty elements," however there is no direct evidence for their involvement in a transfer. Next, one of the copies of the gene was moved to the nucleus and inserted in such a way that transcription would yield the original product. At this point, the product would be produced redundantly by both the mitochondria and the nucleus. Then over the course of many years spontaneous (background level) mutations occurred, some of which rendered the mitochondrial gene copy nonfunctional (turning it into a pseudogene). Had there not been another copy of this gene in the nucleus, such an event would have been lethal. In this case it just further cemented the mutual dependence of the mitochondria and nucleus in coding for certain essential products.

GENE TRANSFER EVENTS FIXED IN THE NUCLEUS

THE PROPOSED scenario for subunit 9 has a directionality, that is, the gene traveled from the mitochondria to the nucleus, and it is the mitochondrial copy that became a pseudogene. But it is important to note the processes of gene movement and inactivation by gene mutation do not have a uniform direction. Viruses and transposable elements are probably indiscriminate in their acquisition of genes (or parts of genes) and transfer to the same genome or to other genomes. This trait is, in part, what makes viruses and transposons such useful tools in genetic engineering. There is, for example, a case in which corn chloroplasts seem to have received a short sequence of DNA from corn mitochondria. There are probably numerous untraceable cases in which genes for single (uncomplexed) products were transferred, but without any evidence remaining—thus making it difficult to know which genome had a particular gene first. Furthermore, gene mutations are generally random, and there should be no guarantee that one or another redundant copy of a gene would be the one that becomes permanently converted to pseudogene status.

Why then does most of the evidence for gene transfer suggest that the usual direction of transfer is from mitochondria (or plastid) to nucleus? Thorsness and Fox (1990) calculated that a mitochondrial to nuclear transfer occurs with a frequency of 2×10^{-5} per replication, while the nucleus to mitochondrial transfer is 100,000 times less frequent. The answer to this imbalance in transfer direction may lie in the differing capabilities of the mitochondrial and nuclear genomes to preserve the transfer event. If one mitochondrion transfers one copy of a gene to a nucleus and if that gene happens to be placed in such a position that it can adequately produce its product without interfering with any other essential gene in the nucleus, then chances are good that transfer event will be passed on to subsequent generations of cells every time the nucleus divides. If, however, a gene is transferred from the nucleus to a mitochondrion or from one mitochondrion to another, it would have far less chance to be passed on to subsequent cell generations.

Although mitochondria replicate along with the nucleus during cell division, there is no guarantee that direct offspring of a particular mitochondrion will be transferred to both daughter cells, whereas all nuclear genes will be. The acquisition of a particular nuclear gene may not confer any distinct advantage to a given mitochondrion, especially if that gene is still functioning in the nucleus. The imbalance in the direction of transfer is thus to be expected.

Which copy of the gene, following a transfer, is more likely to be converted to a pseudogene is also random but more likely to be preserved in such a way that the nucleus retains the good copy. Mitochondrial DNA accumulates mutations nearly ten times faster than does nuclear DNA, and the mitochondrial copy of a shared gene will generally be more quickly inactivated. There may be a selection pressure on the cell as a whole system for the retention of the good copy in a centralized location almost guaranteed to be passed on to offspring rather than in the mitochondrial genome, which is effectively polyploid, and in which any copy of a particular gene has less relative importance. The acquisition of a useful gene by the nucleus may confer a selection advantage to the cell, such that subsequent loss (by conversion to a pseudogene) might be a selective disadvantage. Conversion of a redundant gene to a pseudogene in a mitochondrion or loss of that gene, on the other hand, might not result in any significant advantage or disadvantage in the protected cell environment in which the nucleus is producing the needed product.

All this is to say that although transfer events and gene loss events are probably random, there is a net directionality such that the nucleus seems to acquire genes, while redundant genes are lost in the organelles. There may be a significant lag time between the acquisition of a copy of a gene by the nucleus and the subsequent loss of the mitochondrial copy. For example, a COXII gene was transferred 60–100 mya from mitochondria to nuclei of flowering plants. However, to this day only one genus of legumes is known to have lost the mitochondrial copy (Nugent and Palmer 1991). The tendency toward centralization (into the nucleus) and loss of redundancy ultimately results in a more efficient, streamlined symbiosis. But this trend may nevertheless occur at a slow enough pace that it can be actually observed in process for some genes in some lineages today.

A crucial level of collaboration between nucleus and organelles remains, however, in cases where multiple genes are involved in the production of a complex with multiple subunits, like the F_1 ATP synthetase. Retention of some control by the mitochondria may also be advantageous to the symbiosis as a whole. That is, it prevents either the cell (and nucleus) and the mitochondria from outgrowing the other. Since both genomes must collaborate in producing some products, neither is likely to multiply so quickly as to leave the other behind. In fact, replicative coordination may be a hallmark of any efficient, mutual symbiosis. Pathogenicity and loss of symbionts may lie at the two extremes, depending upon which party outgrows the other.

THE GENETIC CODE IN MITOCHONDRIA

IT IS unlikely that most mitochondria and nuclei are conducting any gene exchanges at the current time, for the simple reason that they no longer speak the same language. At some point in the history of the mitochondria-cell symbiosis, a series of genetic alterations were retained in mitochondria such that parts of their genetic code changed. The series of postulated events leading to this kind of incommensurability is the topic of this section.

The near universality of the genetic code suggests the intuitive conclusion that mutations changing the code are mostly lethal. Such lethal mutations would surely include changes in the genes for tRNAs (transfer RNAs), which are the code translators in all cells. Each tRNA has a site at one end on which a specific amino acid may be attached and a loop containing an "anti-codon" sequence, which complements a specific mRNA sequence, the "codon." Thus the tRNA for leucine should have an anticodon sequence of GAA to complement the codon on the mRNA CUU. A mutation changing the amino acid attachment or recognition loops or the anticodon loop might then change the translation of the genetic code for leucine. However, such a change is very likely to be lethal, because nearly every protein manufactured in the mutant cell would have an incorrect leucine substitution, which would in many cases affect the functions of the proteins. Such a mutant would be quickly outcompeted by any individual capable of making fully functional (nonsubstituted) proteins.

The exceptions to the universal genetic code, of which the mitochondria are a prominent example, must somehow have bypassed the nearly certain lethality of a code change. How did they do it?

The mitochondria-cell association is a highly buffered one in that a tRNA mutation to an individual mitochondrion might result in almost immediate lethality, but the loss of one mitochondrion would usually not confer a net disadvantage to the cell. (In the case of yeast with one large branching mitochondrion, loss of one mitochondrion would be the equivalent of loss of one branch with its particular mutant copy of the DNA.) That newly deficient mitochondrion might, however, not be lost at all, given its ability to use products supplied by the nucleus and given the relative lack of proteins produced exclusively by the mitochondrial genome. The number of mitochondria found in most cells gives these organelles considerable opportunities to experiment, even with ordinarily lethal mutations. The ability to change genetic codes, rare as it is, may be something more easily done by symbiotic individuals protected in large numbers within a host cell and which code for very few of their own genes.

The arrangement of the mitochondrial genome, with tRNA genes alternating with genes for various structural proteins and enzymes, may be the key to the code changes involving punctuation. Obar and Green (1985) noted that the problem with mutations that alter stop codons is that run-on products (like run-on sentences) may be produced from several genes arranged in tandem. However, mitochondrial genomes have few genes coding for proteins in tandem. Many genes alternate with tRNA genes. The way in which newly made tRNAs are clipped off of the RNA transcripts may serve as a sort of stop signal for the genes of proteins (figure 6.1). Such a system would tolerate changes in stop most easily, since the universal system of stops would be dispensable. This kind of change is, in fact, evident in the mitochondrial code. The most consistent alteration has been the change from UGA meaning "stop" to UGA meaning tryptophan (table 6.1).

A process in which RNA transcripts are edited has been discovered in several types of mitochondria. In at least one organism, the plant *Oenothera,* this editing has the net effect of converting the mitchondrial code back to the universal code, by switching a U into a C in the mRNA (Schuster et al. 1990). The details of this editing process are still unclear, but such editing does appear in the case of this plant to be compensation for the code change.

Relatively few mitochondrial genomes have been analyzed, so it is not known how widespread different mitochondrial codes are. The streamlining of the mitochondrial genome by reducing the number of tRNA genes and their use as punctuation seems to have been a

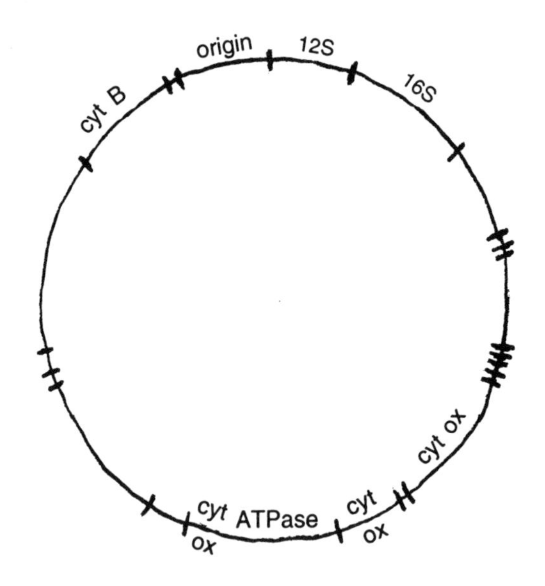

FIGURE 6.1. This is a human mitochondrial genome; tRNA genes (black hashmarks) can act as "punctuation" for other mitochondrial genes, precluding a need for the usual punctuations encoded by nucleotides.

TABLE 6.1. Exceptions in mitochondria to universal cytoplasmic genetic code.

Codon	*Cytoplasmic Code*	*Mammalian Mitochondria (Human)*	*Yeast Mitochondria (Saccharomyces cerevisiae)*
UGA	Stop (opal)	Trp	Trp
AGA	Arg	Stop	Arg
AGG	Arg	Stop	Arg
AUA	Ile	Met (initiate)	Met
AUU	Ile	Ile (initiate)	Ile
CUA	Leu	Leu	Thr

SOURCE: from Alberts et al. 1983.

general phenomenon, and this may be the major selection pressure toward "consolidation" of the genetic code. *Chlamydomonas* mitochondria appear to have retained the universal code (Gellisen and Michaelis 1987). The prediction in the case of mitochondria that still share portions of their codes with the nucleus is that gene transfer may be still ongoing or at least recent. In fact Boer and Gray (1988) determined that *Chlamydomonas* mitochondria may have recently picked up genes for a reverse transcriptase enzyme.

THE STRUCTURE OF MITOCHONDRIA

ALTHOUGH A polyphyletic origin (involving more than one ancestral type) for mitochondria has been hypothesized as a result of diverse morphologies, habitats, and even genetic codes, this has not been supported from 5S rRNA data. Mitochondria, regardless of the form of their cristae (tubular, vesicular, lamellar), are all closely related to the Purple Bacteria. Furthermore, as already discussed, the family tree of early protoctists seems to indicate a single acquisition of mitochondrial symbionts, followed in rare cases by secondary losses.

There is an enormous diversity of shapes, sizes, and numbers for mitochondria. In mammalian cells such as liver cells there may be a thousand elongated, tubelike mitochondria per cell. In some yeasts there seems to be one giant, branching mitochondrion during at least part of the cell cycle. When such a mitochondrion is sliced into multiple sections and examined, it can appear to be many mitochondria. In fact it is only through careful reconstruction of serial microscopic sections that the existence of one large mitochondrion has been demonstrated for yeast (Hoffman and Avers 1973). Mitochondrial shape can be particularly difficult to identify, owing to the tendencies of mitochondria to fuse with each other and to divide. Individual mitochondria or branches of a mitochondrion are approximately bacteria-size, one micrometer in diameter, and are therefore difficult to see under the light microscope even if prepared with the traditional Janus green stain. Mitochondria are visualized best with either a fluorescent dye or in stained sections under the transmission electron microscope.

It is the transmission electron microscope that has revealed the most about the internal structures (ultrastructure) of mitochondria. The typical section reveals an outer membrane enclosing the mitochondrion with a fairly smooth and regular surface. In close proximity is the highly convoluted inner membrane, with its extensive surface area folded to form the internal membrane structures (cristae) (figures 6.2 and 6.3). There is considerable variation among organisms in respect to the shape and number of cristae; the significance of the variation is mostly unknown. Between the outer and inner membranes is the intermembrane space and enclosed by the convolutions of the inner membrane is the matrix space.

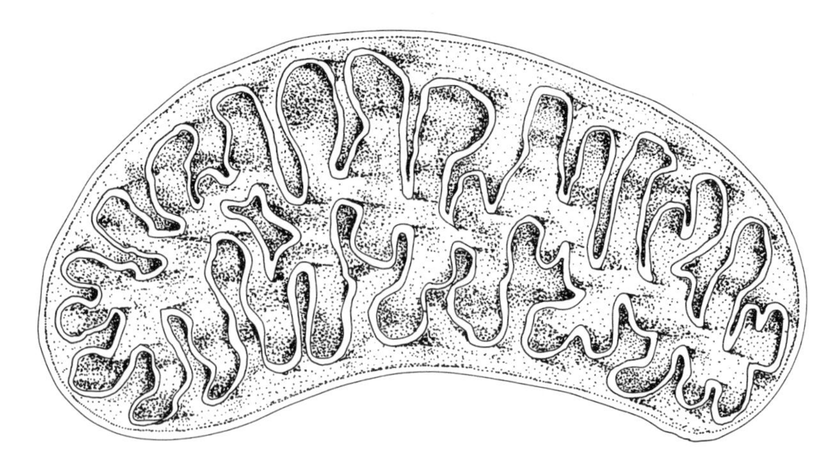

FIGURE 6.2. A mitochondrion cross section. Drawing by C. Nichols.

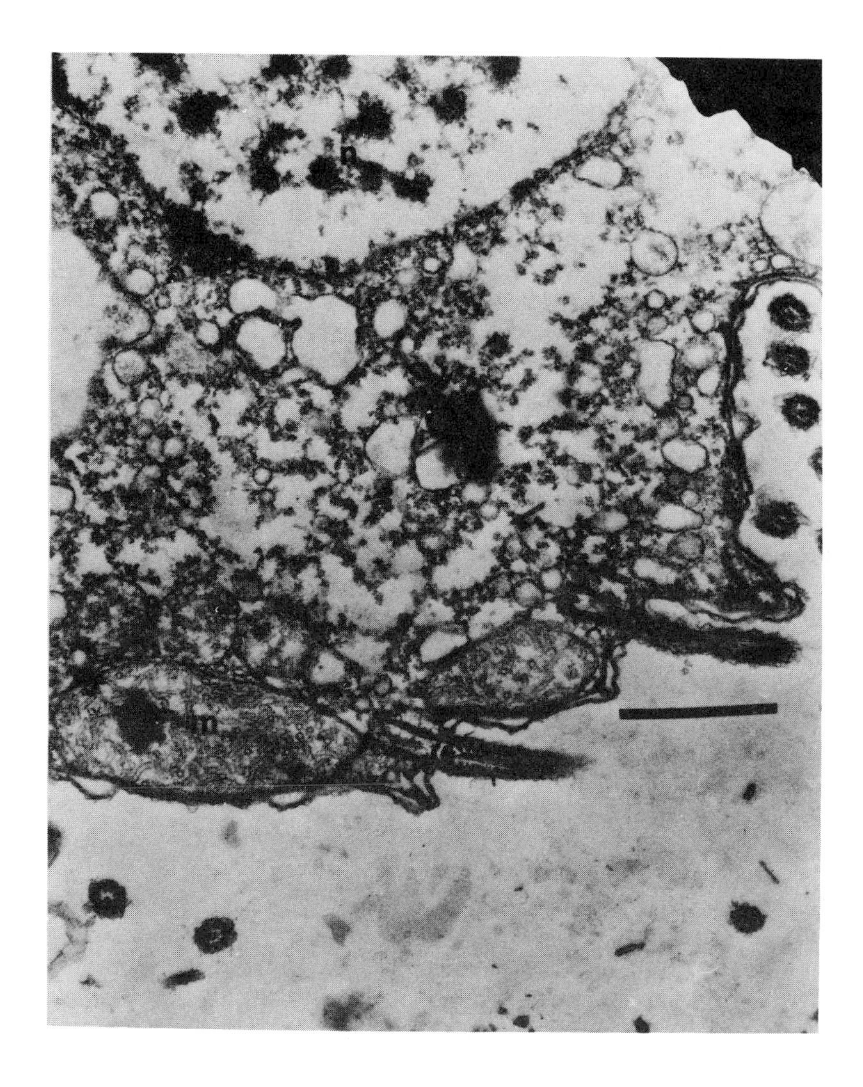

FIGURE 6.3. Mitochondria (M) of a ciliate *Cyclidium borrori*; Nucleus (N); also visible is a 9+0 structure at the base of a motility organelle. Reprinted from Dyer 1989c.

PARACOCCUS AS A CLOSE RELATIVE OF MITOCHONDRIA

CONSIDERABLE EVIDENCE has accumulated to suggest that *Paracoccus denitrificans* (of the Purple Bacteria) is a likely candidate for close descendent of the first mitochondrial symbionts. *Paracoccus* shares a very similar metabolism with mitochondria and perhaps more importantly has strikingly similar ribosomal RNAs.

First, mitochondria have direct bearing on the metabolism of their symbiont partners. The primary function of mitochondria is to process energy from food molecules taken up by the cell and to convert this energy to ATP, a form that can be stored and then instantly put to use. This process is called aerobic respiration (figure 6.4). We have chosen to present a fairly detailed description of aerobic respiration, as it constitutes an important similarity linking mitochondria with their presumed relatives *Paracoccus*. The general features of aerobic respiration in eukaryotes begin outside of the mitochondria, in the cytoplasm of the cell. There, food molecules such as glucose are broken down by fermentation (glycolysis), some of the energy is stored as ATP, and the end product pyruvate is produced. Pyruvate is taken up by the mitochondria and channeled into a pathway called the Krebs cycle in the matrix space (the space in the center of the mitochondrion). In the matrix it is completely oxidized, producing carbon dioxide as a waste product.

During the Krebs cycle, energy-rich reduced molecules NADH (nicotinamide adenine dinucleotide) and $FADH_2$ (flavin adenine dinucleotide) are formed. These two products are channeled into a series of steps called electron transport, which takes place in the internal membranes (cristae) of the mitochondria. The two molecules (NADH and $FADH_2$) transfer electrons to the electron transport chain at two different points. The electron transport molecules are arranged such that complexes of molecules that can pick up both protons (H^+) and electrons (e^-) alternate with complexes that pick up electrons only. As electrons are transported from one carrier complex to another, protons are alternately picked up and released, depending on the capability of the particular complex. The pickups occur on the matrix side of the inner membrane, starting with protons removed from NADH and $FADH_2$ (leaving them in the form of NAD and FAD). Other protons from the matrix space are also picked up by electron transport complexes. The release of protons occurs

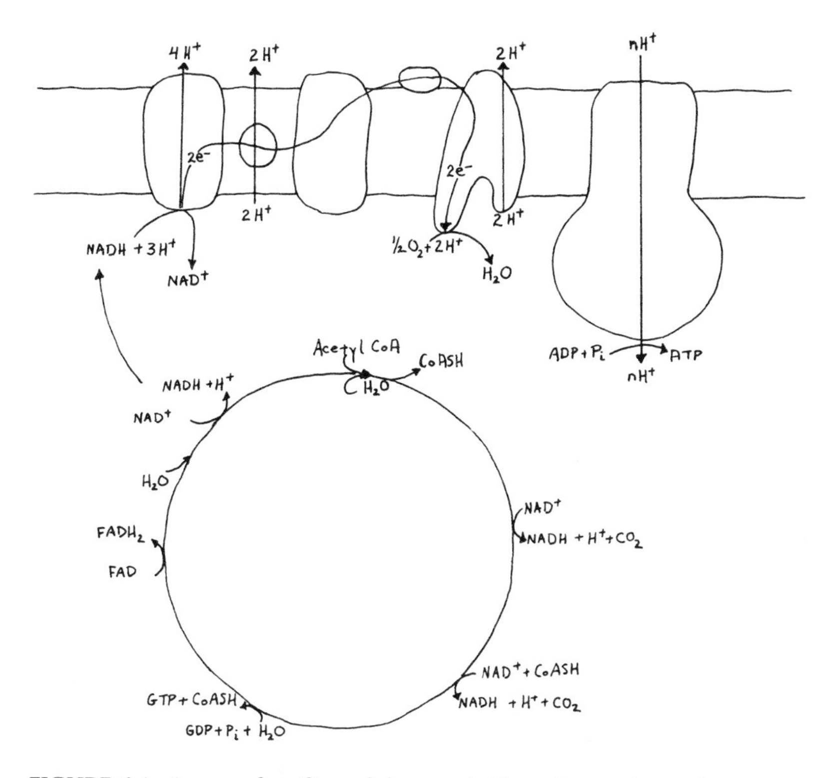

FIGURE 6.4. A general outline of the metabolic pathway of most heterotrophic eukaryotes. Glycolysis (fermentation) takes place in the cytoplasm. Beginning with the Krebs cycle (in the matrix space) shown in part at the bottom, acetyl coA, part of the product of fermentation enters the cycle. NADH and $FADH_2$ are produced (along with CO_2 and a GTP). The electron transport scheme at the top shows an NADH entering the NADH-CoQ reductase complex and releasing its two electrons and a proton (H). As the electrons travel through this and the next four carriers—CoQ, CoQ-cytochrome C reductase complex, cytochrome C, and cytochrome C oxidase complex—more protons are pumped into the intramembrane space. Water is a waste product and terminal acceptor of the two electrons. The protons travel back to the matrix via the ATPase, producing ATP. After Darnell et al. 1990.

into the intermembrane space (the space between the two mitochondrial membranes), such that this area accumulates protons. In a sense the electron transport carriers pump protons across the inner membrane. In order to bring a transport sequence to an end, the electrons that have been traveling from carrier to carrier must be removed. This is accomplished by cytochrome oxidase picking up one final proton for every two electrons and with oxygen forming a waste product, water.

Meanwhile the protons accumulating within the intermembrane space have only one way back into the matrix space and that is through a large enzyme complex called an ATP synthetase. As protons travel through the chain of reactions, the enzyme forms ATP from a precursor ADP and phosphate; water is produced as a waste product. The process is called oxidative phosphorylation.

Specific components of mitochondrial respiration and oxidative phosphorylation (synthesis of ATP) may be found to varying degrees in aerobic bacteria. *Paracoccus,* however, has more mitochondrial features used in mitochondrial fashion than does any other aerobe. John and Whatley (1975) were the first to review the evidence. *Paracoccus* has the components of the Krebs cycle and produces NADH from that cycle, which is then taken up by electron transport. On its electron transport chain *Paracoccus* has flavoproteins, iron sulfur proteins, ubiquinone-10, two b-type cytochromes, two c-type cytochromes, and an a-type cytochrome—all similar to those of mitochondria. Like mitochondria *Paracoccus* is sensitive to electron transport inhibitors like antimycin A and rotenone. Oxidative phosphorylation in *Paracoccus* involves ATP synthetase and seems to be carried out in a mitochondrial fashion. Other similarities of *Paracoccus* and mitochondria include the presence of phosphatidyl choline as a major constituent of the membrane phospholipids.

Dissimilarities between mitochondria and *Paracoccus* seem to be due to the differences between a highly integrated symbiotic association and a free-living habit. For example, free-living bacteria like *Paracoccus* have transport systems necessary for their direct interaction with the environment. *Paracoccus* lacks the extensively invaginated inner membrane of mitochondria. Such membranes in mitochondria may reflect a specialization for generating large quantities of ATP that would be advantageous for the symbiotic association. There are topological differences as well.

The first step in converting the energy of food molecules to ATP, glycolysis, occurs in the cytoplasm of the *Paracoccus,* which is equivalent topologically to the matrix (internal) space of mitochondria. The fermentation products are further processed by Krebs cycle enzymes, also within the cytoplasm of the bacteria. Krebs cycle products NADH and $FADH_2$ enter an electron transport sequence, situated in the plasma (outer, enclosing) membrane; in this case protons are pumped to the outside environment. The ATP synthetases are oriented in the bacterial plasma membrane such that protons enter the cell again producing ATP (figure 6.5).

In addition to metabolic similarities, *Paracoccus* and mitochondria also have similar ribosomes. The most highly conserved nucleotide sequences in cells are those of the ribosomes (rRNA). Ribosomal RNA is, therefore, a useful indication for relatedness. On the basis of rRNA sequences and other criteria, *Paracoccus* is classified as part of the group of Purple Bacteria (Delihas and Fox 1987). Some purples, such as *Rhodopseudomonas spheroides,* when grown in the dark, have a mitochondrial type of respiration. Its inner membrane is highly invaginated with folds and vesicles, as are the membranes of most members of the group. The Purple Bacteria are most similar of all tested bacteria to mitochondria in respect to rRNA. Because the group is almost entirely photosynthetic (but facultatively heterotrophic), this raises questions about what the original mitochondrial symbionts might have been doing. John (1987) suggests that both *Paracoccus* and mitochondria may have retained the highly conserved 5S rRNA, thus providing a signature of their ancestry, while they secondarily lost the photosynthetic aspects of their metabolism. The highly convoluted inner membrane of most purple bacteria might have been retained in the case of mitochondria but not in

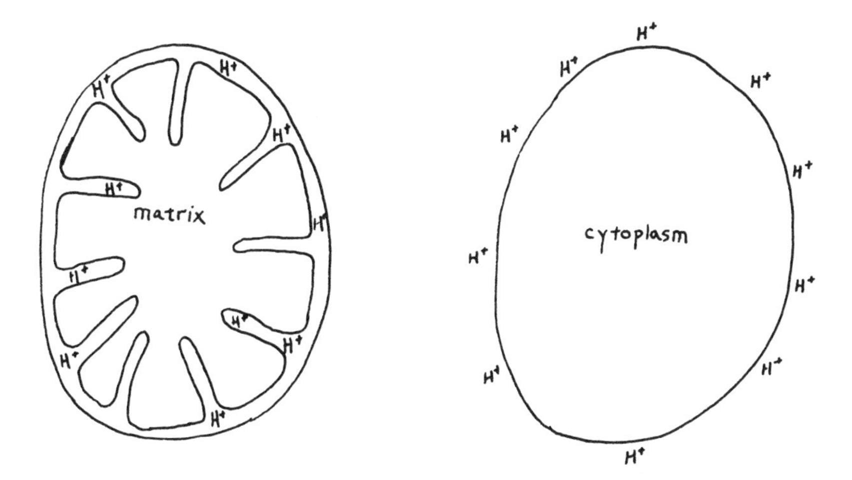

FIGURE 6.5. Topological differences between *Paracoccus* and mitochondria; note that *Paracoccus* pumps protons into the environment while mitochondria pump protons into the space between the membranes.

Paracoccus. The sequence of cytochrome c is also considered to be highly conserved and this too bears out the relatedness of *Paracoccus* to other purple bacteria and mitochondria.

MITOCHONDRIA AND SEX

MITOCHONDRIA HAVE a dynamic spatial relationship with each other within a cell. They divide, bud off, fuse, and change shape. When two mitochondria fuse, there is an opportunity for recombination to occur between the two genomes. Such recombination has been demonstrated experimentally with yeast, one of the organisms of choice in mitochondria research because it is facultatively aerobic. That is, yeast may acquire lethal genetic defects because of mutations or recombinations between mitochondria, but the overall effect need not be lethal to the yeast, which can simply switch to anaerobic metabolism.

For example, yeast mitochondria carrying genes for resistance to chloramphenicol but sensitivity to erythromycin may be "mated" with mitochondria that are sensitive to chloramphenicol and resistant to erythromycin. This may be accomplished by mating two yeast cells with opposite sensitivities; the two haploid yeast cells fuse forming a diploid. After meiosis of the diploid yeast cell, the four offspring yeast cells will be carrying mitochondria—some of which, also, would have fused and recombined their genomes (figure 6.6). Most of the mitochondria should have the same genes they began with; a few will be sensitive to both antibiotics and a few will be resistant to both antibiotics. These are the "recombinants" and they are evidence that mitochondrial fusions can result in genetic recombinations (Gillham 1978).

There are technical problems in this type of recombination experiment (Gillham 1978). One is that a mixed population of mitochondria, with some individuals retaining the original complement of genes and others being recombinants, might be entirely contained within one yeast cell, which would grow ambiguously on the various selection media. A solution to this problem is to grow the yeast rapidly through many (about twenty) divisions on nonselective media and then to attempt to grow the many offspring on selection medium. This process increases the chances of having some yeast offspring that are fairly homogeneous in respect to their mitochon-

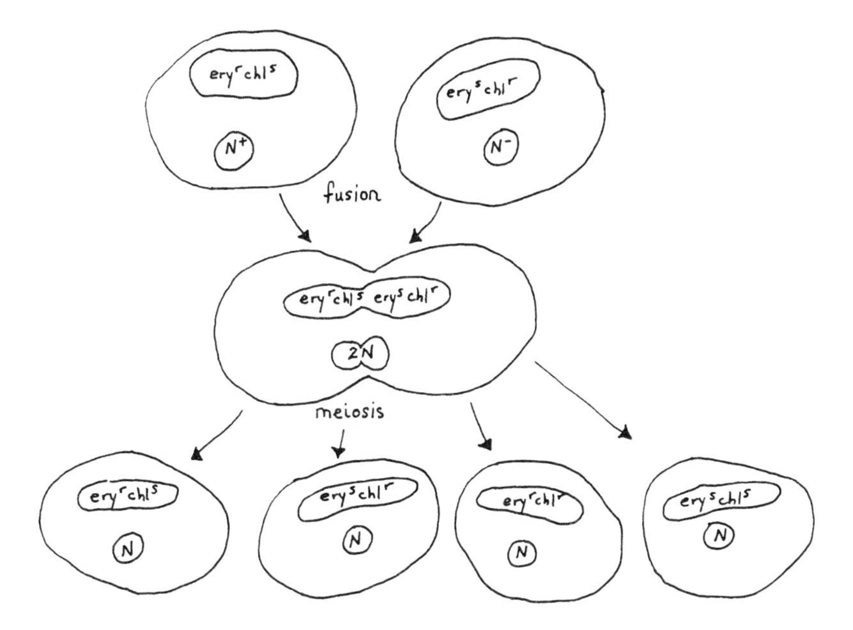

FIGURE 6.6. Yeast mating including the mating outcomes of two mitochondrial mutants. Labels ery^r = erythromycin resistant; str^r = streptomycin resistant; s = sensitive. Note that most offspring will have had little or no recombinations, but a few will have recombined and may be selected for on special medium.

dria just by chance dilutions of the original heterogeneous population. Alternatively buds of the yeast cell may be isolated, grown, and tested on selection medium. The chances are good that a particular bud will be homogeneous in respect to mitochondria.

There is also a conceptual problem if mitochondria recombination experiments are conducted using the yeast *Saccharomyces* and its single giant branching mitochondrion. It is apparent from these sorts of experiments that the individual branches are known to act, in effect, like individual mitochondria. Thus it seems that one branch can recombine with a branch from a mitochondrion in another yeast during fusion, while retaining to some extent a genetic identity unique to that branch. This would be true only up to a point because the lack of membrane barriers between genomes of individual branches probably results in considerable recombinations within the mitochondrion.

The occurrence and importance of mitochondrial recombinations in nonexperimental systems are not well understood. However, it is likely that such recombinations would be an efficient way to fix an advantageous mutation in a population of mitochondria. Recombination could also serve to buffer a population of mitochondria or a large single mitochondrion from lethal effects of mutations. Mitochondrial recombinations within a homogeneous population would have no net effect. Mitochondrial recombinations following fusion of two host cells during sex might be an effective way to both spread and buffer mitochondrial mutations through a cell population.

LEAKY MATERNAL INHERITANCE

UNLIKE YEASTS, not all cells pass on their mitochondria during sexual fusions. Mammals, for example, are constrained by the size of the male gametes (sperm) that must penetrate an egg. Each sperm carries only one mitochondrion, sufficient for the sperm cell's own metabolic needs, (each mitochondrion with 50 to 100 copies of the mitochondrial genome). But this mitochondrion usually does not survive the fusion. By contrast, the egg has 10^5 to 10^8 copies of the mitochondrial genome. Therefore most mammalian mitochondria are inherited maternally—passed exclusively in the large volume of egg cytoplasm found in mammals and in many other organisms.

The exceptions to maternal descent of mitochondria are, however, notable. About 1 in 10,000 mitochondrial genomes of mice are paternal, the sperm mitochondrion having survived fusion, then gaining a presence in the developing organism (Gyllensten et al. 1991). Paternal mitochondria have an opportunity to fuse and recombine with maternal mitochondria. It would be expected then that mitochondrial genomes are to some extent heterogeneous, even in organisms with a predominant pattern of "maternal inheritance." And given the opportunities for fusion, mitochondrial genomes may be more realistically considered chimeras of maternal and paternal sequences.[1]

[1] Excitement was generated several years ago in the popular press that Eve's mitochondrial genes might be extrapolated from extant sequences. Unfortunately, even 1 in 10,000 paternal mitochondria leaking in would be sufficient to make modern mitochondrial genomes mixtures of both Adam and Eve.

HYDROGENOSOMES: DEGENERATE MITOCHONDRIA?

HYDROGENOSOMES ARE membrane-bounded organelles involved in the anaerobic or micro-aerophilic metabolism of some of the ciliates and all of the parabasalids, a group that includes *Trichomonas*. Hydrogenosomes are defined in part by their hydrogenase (H_2 generating) activity, part of a fermentative pathway in which pyruvate is oxidized to H_2, CO_2, acetate, lactate, and other products (Muller 1988). The metabolism of hydrogenosomes has some similarities to that of mitochondria. Both convert pyruvate to acetyl coenzyme A, and both have some of the enzymes of the Krebs cycle (Muller 1988). Most other metabolic similarities seem to have been secondarily lost, or were never there to begin with. Some (and perhaps all) hydrogenosomes have double membranes although, unlike in mitochondria and plastids, the inner membrane is usually not enlarged and folded. Exceptions in which the inner membrane is folded include the ciliate *Metopus contortus* (Finlay and Fenchel 1989).

Hydrogenosomes have not been found to contain DNA or any sort of independent genetic system. Nevertheless, several authors hypothesize that these organelles are of symbiotic origin. Muller (1988) suggests that a fermentative bacterium similar to *Clostridium* may be a relative of hydrogenosomes. Others (e.g., Finlay and Fenchel 1989) consider hydrogenosomes to be degenerate mitochondria.

Several observations support the hypothesis of hydrogenosomes as degenerate mitochondria. The fact that hydrogenosomes are found in some ciliates, a recent group on the evolutionary tree and one in which most members have fully functional mitochondria, suggests that hydrogenosomes may have been secondarily derived when certain genera became anaerobic. These ciliates include such disparate groups as symbionts in the rumens (forestomachs) of cud-chewing mammals as well as free-living inhabitants of anaerobic sulfur-rich environments. Thus the degeneration of mitochondria into organelles such as hydrogenosomes may well be polyphyletic and a frequent event in the evolution of secondary anaerobes. The double membranes of hydrogenosomes, which in some ciliates include convolutions of the inner membrane, also resemble those of mitochondria.

Turning now to the second group in which hydrogenosomes are present, the parabasalids are distinctive in that they are all symbionts living in anaerobic habitats, such as intestinal tracts of animals. The parabasalids include no completely aerobic members. Thus the entire group might be defined by a secondary loss of aerobic function. On the other hand, perhaps the parabasalids never did acquire mitochondria. Hydrogenosomes in this group would then constitute a different lineage of symbionts from those of ciliates or mitochondria in general.

Some free-living anaerobic ciliates have symbiotic bacteria associated with their hydrogenosomes. (This is similar to the symbiosis of *Methanobacillus omelianskii* described earlier in this chapter.) These bacterial symbionts are methanogens, which collect waste products of the hydrogenosomes, H_2 and CO_2, for use in their own metabolism.

Most methanogens are autotrophs, and these symbiotic varieties may be releasing some fixed carbon to their host. Also the host may digest some of the symbionts. In some ciliates such as *Plagiopyla* the hydrogenosomes alternate with methanogens in a stacked dish or coin formation, and the two divide synchronously in coordination with the host ciliate so that constant numbers are maintained—about 3,500 bacteria and 5000 hydrogenosomes per cell (Fenchel and Findlay 1990a, 1990b).

LACK OF DNA IN HYDROGENOSOMES

THE SYMBIONT-LIKE division of hydrogenosomes in tandem with the host ciliate is not, however, attributable to specific DNA retained in the organelle of the putative symbiont. If the hypothesis of symbiogenesis is correct, all of the original DNA must have been lost or transferred to the nucleus. Although DNA and a semiautonomous genetic system are important criteria for establishing an organelle as a symbiont, there may be some circumstances, such as an extreme secondary loss of function in a symbiont, in which most all traces of DNA are also lost or transferred. Hydrogenosomes are (fortunately for this argument) not the sole example. In chapter 7, plastids of parasitic (nonphotosynthesizing) plants will be seen to have experienced a precipitous loss of DNA. In chapter 4 recall, microbodies—single membrane-bound organelles that appear to divide and yet have no DNA—were also discussed as possible symbionts. Finally, motility organelles (chapter 8), which either have the bare remnants of a genome or perhaps none at all, will be analyzed for clues to a symbiotic origin.

In fact, the size of a symbiont genome may have a direct relationship with the number of unique and necessary functions that the symbiont supplied to the host during the initial stages of the relationship and with the number of secondary losses (table 6.2). Therefore, the robust genomes of green plastids may be a consequence of the uniqueness and necessity of photosynthesis. The scant or nonexistent genomes of plastids in parasitic plants, the hydrogenosomes of ciliates and parabasalids, and perhaps motility organelles in a wide range of eukaryotes may all be a consequence of the sudden lack of necessity of most of the primary functions of the organelle.

TABLE 6.2. Loss of DNA in symbionts.

Organelles in Order of Genome Size, (from large to undetectable)	Major Function Supplied	Redundancy with Host	2° Loss of Function
Plastids	photosynthesis	? none	
Mitochondria	Fermentation, respiration	Fermentation	Fermentation
Plastids of parasitic plants	? none	? none	Photosynthesis
Motility organelles	Tubulin fermentation motility	Fermentation, some types of motility	Fermentation, some types of motility
Hydrogenosomes	Anaerobic metabolism	? none	Respiration *
Peroxisomes	Oxygen detoxification and other specialized functions	? none	?

* or if hydrogenosomes are symbionts of a different lineage perhaps they never had respiration.

THE ACQUISITION OF MITOCHONDRIA

THE EVIDENCE for a symbiotic origin for mitochondria is both compelling and well accepted by most biologists. We began the chapter with the suggestion that acidic waste problems might have created the selection pressure for the acquisition of a respiring symbiont (essentially an acid eater). In the remainder of the chapter we addressed the six points that firmly establish an organelle as a symbiont. The presence of a genome in mitochondria is a direct piece of evidence. Nevertheless, we spent many pages explaining just how complicated that little genome is. It shares the coding for some gene products with the nucleus (the cement that binds them) and has lost many other genes. The genetic code in mitochondria is somewhat different from the nearly universal genetic code. How code differences might have evolved and what the consequences may be are not well understood, but we attempted an explanation.

Paracoccus is a likely relative of mitochondria, sharing many aspects of the rather complex process of aerobic respiration (described in detail). The fact that mitochondria still maintain a type of semi-independent genetics, in spite of a greatly reduced genome, means that eukaryotic genetics is really quite complicated. Not only does it involve the genetics of the nucleus but also a whole population of breeding symbionts.

Finally we addressed one difficult problem: the origin of another organelle and possible degenerate mitochondrion, the hydrogenosome. In contrast to the mitochondrial story, the apparent lack of a genome in hydrogenosomes makes interpretation difficult. This brings up the general problem of how to identify putative symbionts when the genome has been reduced to nothing. Few examples are so clearly established as those of mitochondria and plastids (in the next chapter). Other more tentative criteria than presence of a genome may be necessary for other eukaryotic structures and organelles.

VII
The Acquisition of Plastids

SOME OF the eukaryotic cells that had acquired mitochondria also acquired additional symbionts, in most cases photosynthetic prokaryotes (cyanobacteria), which became the plastids. The identification of plastids as former symbiotic prokaryotes will follow a similar set of criteria to that used in the previous chapter on mitochondria (see also appendix B).

CULTIVATING A FOOD MAKER

THE QUESTION of why a heterotrophic host cell might have an advantage in acquiring a photosynthetic symbiont has an easy answer. This is a symbiotic association that has happened over and over again in many heterotrophic groups of organisms. In the animal kingdom at least five phyla have members with photosynthetic symbionts: coelenterates (jellyfish, coral, hydra), molluscs (clams, sea slugs), platyhelminths (flat worms), poriferans (sponges), and urochordates (sea squirts). All these hosts of photosynthetic symbionts have a distinct advantage of an easy supply of food from their symbionts as long as they are willing to sunbathe in the photic zone.

Many protoctists are photosynthetic, with their own plastids acquired long ago. However, some of the heterotrophic protoctistan phyla have members with photosynthetic symbionts. These include ciliates, actinopods, and granuloreticulosans (foraminiferans). The fungi are, of course, famous for their associations with algae in lichen symbioses; about 15,000 species of ascomycete fungi participate in lichens. Fungi also form symbioses (as mycorrhizae) with the roots of many plants. There, in exchange for sugars provided by the host, the fungi facilitate root uptake of vital soil minerals.

Many photosynthetic symbionts are just a step away from digestion—and their hosts just a step away from herbivory. There seems to be a complex set of interactions between host and symbionts that prevent such hostilities from arising under most circumstances (as long as there are plenty of nutrients and light). Muscatine and Pool (1979) studied symbioses of hydra, paramecium, and coral and reported a remarkably stable number of algal symbionts per host cell under constant light and nutrient conditions. The regulatory mechanisms include expulsion of moribund or dead cells, digestion of cells, and inhibition of algal growth with growth inhibitors or by limiting nutrients. The algae seem to have mechanisms to prevent digestion, such as the production of inhibitors and resistant envelopes and movement away from digestion areas.

A sound hypothesis for the acquisition of the original plastid symbiont (which may have occurred independently more than once) is that the requisite evolutionary steps followed the sequence readily observed today in so many of the symbioses mentioned above. According to rRNA data plastid acquisition probably occurred shortly (geologically speaking) after mitochondria were established. The fact that all plastid-bearing cells have mitochondria strongly suggests that mitochondria preceded plastids. However, this may just indicate that mitochondria are an obligate accessory for the use and maintenance of plastids. In any case, microtubule and other cytoskeletal systems were probably well established by this time, having preceded both mitochondria and plastids.

The mechanism for symbiont uptake may have been as simple as ingestion by phagocytosis and retention in food vacuoles. Symbionts resistant to digestion and/or their uptake by hosts with less efficient digestion mechanisms would start the association in the direction of symbiosis. Photosynthesis in sunny, nutrient-rich environments tends to be overproductive to the point of being "leaky." This high photosynthetic potential, coupled with evolutionary devices such as toxicity, is partly why grazing herbivores in general do not usually wipe out entire populations of photosynthesizing individuals. The abundance of photosynthates—in the form of sugars, leaves, fruits, etc.—often provide easily for the continued growth of the photosynthesizer itself as well as for the grazers. The ability of many photosynthesizers to provide for more than just their own bare needs makes them aptly suited for the role of endosymbiont (internal symbiont).

In turn, the advantage to the photosynthetic symbiont would be protection in a motile, culture chamber that is somewhat safer from other herbivores and protected from desiccation and excess ultraviolet light. It is probably no coincidence that weakly motile or nonmotile algae and cyanobacteria often form symbioses with highly motile protoctists and animals. Symbionts that by chance release more of the extra photosynthate to the host would enhance the symbiosis, and the pair itself would be selected for. Steps by which the association might become more obligate would include mechanisms that would virtually guarantee that the relationship would not be broken. This, in the history of plastids, included the ultimate step: genetic integration.

THE GENETICS OF PLASTIDS

PLASTIDS CONTAIN circular molecules of DNA ranging in size from about 80 to 275,000 base pairs. The only known exception to circular DNA in plastids is that of the green alga *Acetabularia,* which has both linear and minicircular DNA and perhaps considerable variability in quantity of DNA. In the few kinds of plastids for which detailed genetic studies have been made, each organelle appears to have numerous copies of DNA (from 13 to 1000 copies) clustered in areas called nucleoids (Dyer 1984). Plastid DNA codes for 2 to 4 rRNAs and about 35 tRNAs for use in the plastid and probably about 100 other gene products (Dyer 1984). During DNA replication and division of the host cell, the plastids also replicate their DNA and divide, assuring transmittal to the next generation of cells. In plant cells devoted to photosynthesis, such as leaf cells, plastids continue to divide after the cell has stopped dividing and growing larger, thereby increasing the total volume devoted to photosynthesis in each cell.

As in the mitochondria considerable gene transfer has occurred from the plastid to the nucleus. Some large enzyme complexes such as ATP synthetase are made up of several subunits, some encoded by the nucleus and some by the plastid. Shared coding, as discussed in the previous chapter, is good evidence for gene transfer events, most of which probably occurred from plastid to nucleus. Mutual genetic control over essential complexes is a likely mechanism by which symbiotic associations are maintained and kept from crossing the line into pathogenicity for either party. That is, the plastids are not able to overrun the host cell because they are limited by the nucleus in respect to how much ATP synthetase and ATP they can make. Likewise, the host cell cannot grow so fast that it dilutes out the symbionts because it too is dependent upon the production of ATP. Other important complexes and pathways that are products of shared coding by the nucleus and plastids are the photosystem I complex and the photosystem II complex (Dyer 1984), which capture light energy for the synthesis of sugars.

In corn, at least two DNA sequences are found in both the mitochondria and the chloroplasts (Stern and Lonsdale 1982; Dyer 1984). This suggests that in cells with both organelles present, such as plants or algae, genetic transfer and perhaps even recombination is possible between species of organelle within the same cell.

EXTREME LOSSES OF PLASTID GENES

PARASITIC PLANTS present the interesting problem of what happens to genes for the various photosynthetic pathways when the products are no longer needed. *Epifagus,* the beech drop, is a flowering plant that is parasitic on beech trees. As with other colorless, nonphotosynthetic plants, *Epifagus* takes its nutrients from other plants that it exploits. The 71,000 bp genome of *Epifagus* lacks approximately 30 genes for photosynthesis and plastid respiration, as well as genes for some tRNAs and RNA polymerase (now provided by the nucleus). Because the parasite seems to have diverged from a free-living green ancestor only 5 to 50 million years ago, *Epifagus* shows that unused genes may be lost rapidly (de Pamphilis and Palmer 1990; Wolf et al. 1992).

Organisms first entering a symbiotic association presumably arrive with all of the genes necessary for a free-living existence. It is not surprising that many of these genes would subsequently be lost, yielding much smaller genomes, as observed in mitochondria and plastids. The example of *Epifagus* suggests that genes are lost quickly, and this implies that the period of establishment of mitochondria and plastids in irreversible, highly integrated relationships may be only a blink of an eye in the fossil record.

But, then, why do chloroplasts of parasitic plants linger? Why do they retain a genome at all? The synthesis of porphyrin (needed by the mitochondria) and of various amino acids are essential functions maintained in perhaps all plastids, and this suggests that the plastid genome can not be entirely lost in plastid-bearing organisms (Howe and Smith 1991; Wallsgrove 1991). Analogous to the situation with parasitic plants are several incidences in the protoctist kingdom in which plastid functions must have been secondarily lost. One fascinating example is the apicomplexans (including the malaria parasites), which rRNA data show to be close relatives of both ciliates and dinoflagellates. Some of these dinoflagellates are photosynthetic, having acquired what appears to be a distinctively eukaryotic photosynthetic symbiont (which will be discussed later). Amazingly, in spite of a long evolution as animal parasites, the apicomplexans still apparently retain a plastid genome (about 35,000 bp) of a dinoflagellate ancestor (Walters 1991; Palmer 1992).

Another example of a genomic vestige that records a plastid symbiogenesis is that of the nonphotosynthetic euglenid *Astasia*, which retains a 61,000 bp plastid genome, inherited from its photosynthetic ancestor (Siemeister and Hachtel 1989).[1] Palmer (1992) muses whether these examples are just the tip of the iceberg and whether other heterotrophs bear nonphotosynthetic plastids as well. Likely candidates might include oomycetes (water molds) and acanthamoebae (spiny amoebae) and perhaps others.

[1] A *Paracoccus* type bacterium, possibly the ancestor of mitochondria, was probably a photosysnthesizer or had recently lost photosynthesis. It is in the group of purple bacteria characterized by presence of or 2° loss of photosynthesis.

STRUCTURE OF PLASTIDS

PLASTIDS COME in a diversity of shapes, including disks and ovals and more unusual shapes like spiraling ribbons or horseshoes (figure 7.1). The size range is usually one to ten micrometers; however some plastids are large enough that those in *Spirogyra* were described by van Leeuwenhoek, the seventeenth century Dutch microscopist, as "green streaks, spirally wound serpent-wise and orderly arranged" (Dobell 1958). Photosynthetic cells may contain just one or up to hundreds of plastids, usually brightly pigmented with specific chlorophylls and accessory pigments. The most familiar and common plastid, the chloroplast of plants and green algae, has, for example, a pigment signature containing the greens of chlorophyll a and chlorophyll b, along with the yellows of carotenoids.

Plastids are usually bounded by two membranes (although there are important exceptions), and they contain an additional set of internal membranes in the form of flattened disklike compartments, the thylakoids. The thylakoids may be present in a stacked format (like an irregular pile of pennies), with connectors between the stacks forming a continuous space (figure 7.1). A stack of thylakoids is termed a granum (plural grana). The area enclosed by the two outer membranes, and in which the grana are suspended, is called the stroma.

The general topology of the plastid is probably analogous to that of mitochondria. The cristae of mitochondria correspond to the thylakoids of plastids. The matrix space of mitochondria corresponds to the stroma of plastids. The major spatial difference between the two types of internal membranes, cristae and thylakoid, is that the thylakoids seem to have pinched off completely from the inner membrane to form a distinctive set of third membranes. The mitochondrial cristae still maintain their connection with the inner membrane in the form of intricate folds and convolutions (figure 7.2).

THE METABOLISM OF PLASTIDS

THE METABOLISM of plastids is complicated. It includes the use of light-harvesting pigments that pass electrons along electron trans-

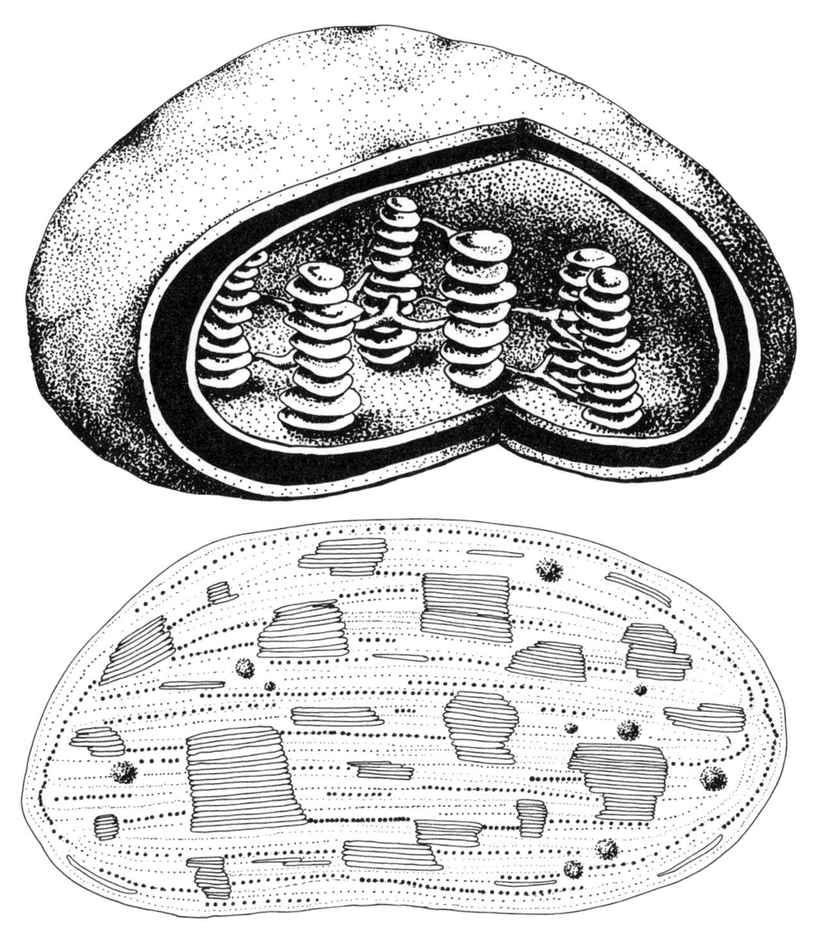

FIGURE 7.1. Cross sections of a plastid showing grana. Drawing by C. Nichols.

port pathways for the purpose of making energy-rich molecules NADPH and ATP. These two molecules are then channeled to the carbon fixation or Calvin cycle. The next several pages are devoted to a detailed description of these processes. This level of detail is necessary, we believe, because the entire complex metabolic process is so strikingly similar to that of the proposed relatives of most plastids, the cyanobacteria. Thus the metabolism of plastids constitutes an important piece of evidence linking them to their symbiotic ancestry.

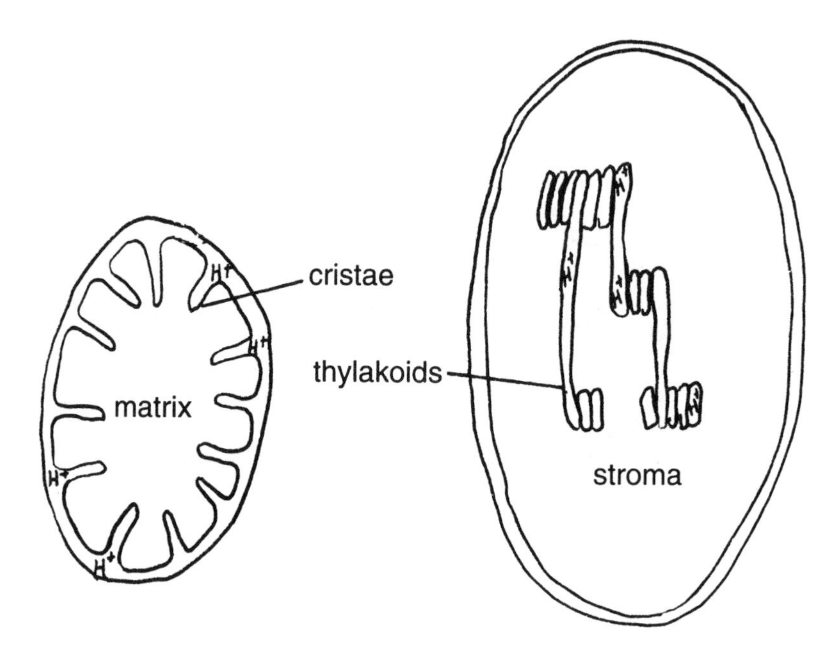

FIGURE 7.2. Comparison of the topology of mitochondria and plastids.

Chlorophylls are the major light-capturing pigments embedded in the thylakoid membranes. They occur in clusters, along with accessory pigments, carotenoids. These clusters are thus light-harvesting complexes. As photons bombard the complexes, electrons in the chlorophyll become excited, and energy is transferred to a particular chlorophyll in the center of the cluster, the reaction center. Electrons become excited (or, in this case, photoexcited) when the discrete "packets" of light energy that constitute photons are absorbed by the pigment molecule—provided that enough energy is absorbed to cause an electron that was traveling in a low orbital around the nucleus of an atom to be transferred to a higher energy orbital. These excited electrons may then transfer energy to a different atom. When this energy (in the form of excited electrons) is channeled into an electron transport sequence, it is on its way to being converted into chemical energy. The reaction center is closely associated with an electron acceptor molecule and an electron donor molecule. When an excited electron from the reaction center is passed to the electron acceptor (to begin its course along an electron transport sequence), the electron donor quickly fills the "hole" with a new electron.

In plastids with both chlorophyll a and b, chlorophyll a may be found throughout the light-harvesting complex, including the reaction center which passes electrons on to the electron transport of what is called photosystem I. Chlorophyll a picks up photons at wavelengths of 700nm; not surprisingly, this wavelength constitutes one of the peaks in the signature of sunlight. Chlorophyll b, on the other hand, is present throughout the light harvesting complex but not in the reaction center. It is excited by photons from the sun at wavelengths of 680 nm (which have more energy than photons at 700 nm). Chlorophyll b transfers its captured energy to the chlorophyll a in reaction centers. Energy from electrons excited in chlorophyll b is channeled via chlorophyll a into the electron transport of what is called photosystem II (figure 7.3).

Accessory pigments such as carotenoids may pick up photons at different wavelengths and pass the energy on to the reaction centers. Thus an efficient chloroplast can take advantage of several wavelengths of light. The specificity of pigments in absorbing wavelengths of light is in part what defines them as pigments. When pigments are bombarded with all wavelengths of light (i.e., white light), they absorb particular wavelengths and reflect the rest. A green pigment, such as chlorophyll, absorbs energy in the red color range; thus the label green distinguishes what this particular molecule reflects, rather than what it absorbs. Because pigments are distinguished by their absorption of various wavelengths, they will almost always look colorful to our eyes.

Because the energy of photons is transferred to the reaction center by way of excited electrons, the system requires a source of electrons as well as a source of photons. Where do these electrons come from? In plastids, the donor molecule is water, which is always in abundant supply in an active cell. But water is a reluctant donor, as it is a fairly stable molecule. An enzyme, called the oxygen evolving complex, when in the presence of light will split water in photosystem II in a process called photolysis.

$$2\,H_2O \rightarrow 2H^+ + 2e^- + \tfrac{1}{2}O_2.$$

Beginning with a water molecule, two electrons (e^-) are transferred to the nearby reaction center. The two protons or hydrogen ions (H^+) are used later and indirectly in forming ATP. The molecular oxygen is given off as a waste product.

Photosystem II, receiving energy from chlorophyll b (channeled into a chlorophyll a at the reaction center) is where the electron

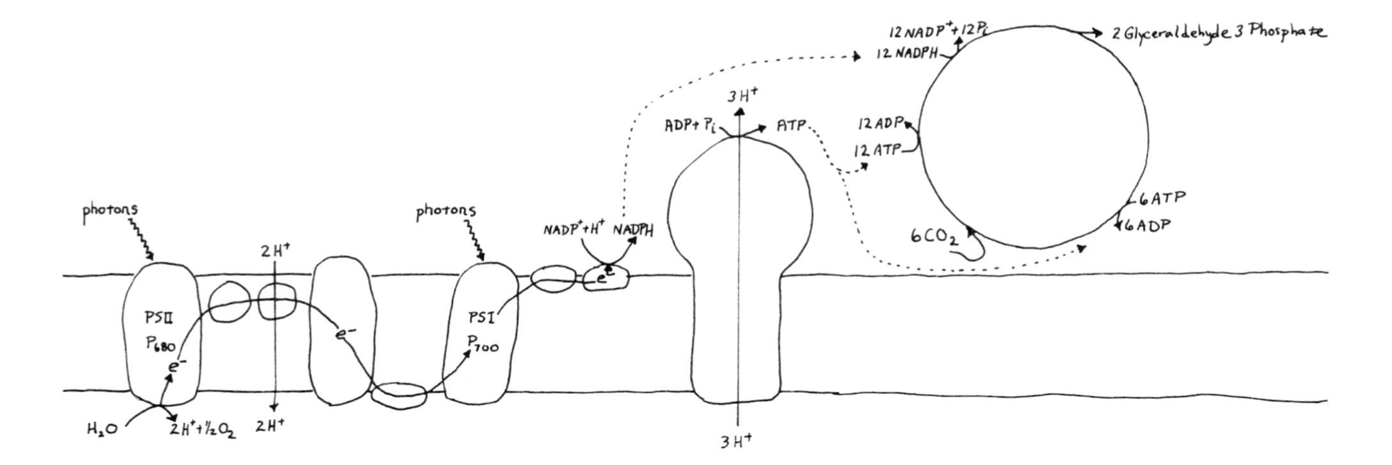

FIGURE 7.3. An ultimate goal of this photosynthetic process is to produce a sugar, one half of which (glyceraldehyde 3 phosphate) is produced on the far right of the diagram. Starting on the left, light causes electrons to begin their journey through a series of carriers. Electrons are constantly being replaced by ones released from the breakdown of water. Beginning at photosystem II (PSII), the electrons then travel through a duo of plastoquinones, a cytochrome b/f complex, a plastocyanin and then to photosystem I (PSI) to receive another stimulating dose of light, and on to a ferredoxin and an NADP reductase. The protons pumped from the stroma to the thylakoid space travel back through the ATPase to make ATP. The ATPs and NADPHs are used in the Calvin cycle to make parts of sugars. Note that two revolutions of the cycle result in two molecules of glyceraldehyde 3 phosphate, thereby using up 18 ATPs and the H^+ from 12 NADPHs. After Darnell 1990.

transport part of photosynthesis begins. Photosystem II gets its name because it appears to have evolved after photosystem I, even though in most diagrams of photosynthesis photosystem II is shown first.

As in the mitochondria, electron transport occurs through a sequence of carrier molecules embedded in the thylakoid (in mitochondria, cristae) membranes. Carriers are able to either pick up and transfer both electrons and protons or electrons only. In photosystem II a plastoquinone is the first carrier, and it seems to be the only one that picks up protons from the stroma space such as the protons from photolysis. The next carrier, a cytochrome b/f complex, receives both electrons and protons, pumping them into the lumen of the thylakoids. After this step electrons are passed to photosystem I. Then, with the help of extra energy passed to the photosystem I reaction center from chlorophyll a, the electrons continue on their journey. After transfer through more carriers, the electrons are picked up by $NADP^{+}$, which with a proton (H^{+}) from the stroma forms an energy storage molecule NADPH. By the end of the electron transport sequence, protons will have accumulated in the lumen of the thylakoids. Just as with the mitochondria, the protons reenter the matrix space via ATP synthetase complexes embedded in the membrane. ADP enriched by a third molecule of phosphate forms energy-rich ATP.

This linear system in which photosystems II and I are linked in one fairly continuous pathway generates relatively few ATPs, although usually a sufficient number of NADPHs for use in the dark reaction (described later). An alternative cyclic pathway may be used to generate more ATPs. In this pathway some of the carriers from photosystem II, the one that generates a proton gradient, are used by photosystem I. At the end of this alternative pathway the electrons are sent back to photosystem I, forming a cycle. The advantage is that additional ATPs may be made (from the additional protons in the lumen), but NADPH is not made. As will be shown in the next section, sugar manufacture occurs in the dark reaction, and it requires more ATP than NADPH.

MAKING SUGAR

THE SYNTHESIS of sugar occurs in the "dark" reaction (or Calvin Cycle) by way of enzymes and intermediates that are soluble in the aqueous medium of the stroma. The major enzyme of sugar synthesis is ribulose 1,5 bisphosphate carboxylase. Indeed, detection of this enzyme is taken as a reliable indication of the presence of the entire Calvin cycle in a cell.

ATP supplies the energy for the reactions in the Calvin Cycle and NADPH supplies the protons. In the expensive set of reactions outlined in figure 7.3, 18 ATPs and 12 NADPHs are used to convert six CO_2 molecules to two glyceraldehyde-3-phosphate molecules. Glyceraldehyde-3-phosphate is then transferred out of the plastid and into the cytoplasm of the cell, where the two molecules are joined to make one glucose molecule. Because light is necessary for the production of ATP and NADPH, the Calvin cycle eventually runs out of energy if cells are kept in the dark.

Ribulose 1,5 bisphosphate carboxylase actually catalyzes two competing reactions. It produces glyceraldehyde-3-phosphate, but it also converts glyceraldehyde-3-phosphate to glycolate in a process called photorespiration. Photorespiration requires oxygen and thus could not have evolved until free oxygen was available. However, photosynthesis itself provides oxygen as a waste product, and therefore photorespiration could have evolved nearly simultaneously.

Although the structure of the active site of the enzyme ribulose 1,5 bisphosphate carboxylase is necessary for the enzyme's function as a synthesizer, it is also fortuitously appropriate for an additional function as an oxygenase. The product, glycolate, is transported out of the chloroplast and into microbodies called glyoxysomes (suspected, themselves, of being derived from symbionts). There the glycolate is oxidized to glyoxylate; then glyoxylate is transported to the mitochondria and converted to glyceraldehyde-3-phosphate. This large molecule is then channeled back to the plastids where it may be put to eventual good use in making glucose (or in photorespiration). This long and complicated process seems to minimize the negative effects of the loss of hard-won glyceraldehyde-3-phosphate molecules effected by the "traitorous" ribulose 1,5 bisphosphate. Some amino acids are also made in this sequence of reactions.

The fact that mitochondria are necessary in this extra cycle fits the observation that there are no cells which have plastids but which lack mitochondria. And this, in turn, supports the idea that mitochondria preceded plastids in the symbiogenesis of eukaryotic cells.

SYNECHOCOCCUS AS A CLOSE RELATIVE OF PLASTIDS

RIBOSOMES ARE universal in cells and are known to be highly conserved. Therefore, specific sequences of genes encoding ribosomal RNA are considered to be useful indicators of the relatedness of cells. On the basis of the sequence for ribosomal RNA genes, a genus of cyanobacteria, *Synechococcus,* appears (of all the bacteria tested) to be most closely related to almost all plastids (Delihas and Fox 1987; Volker et al. 1987). This result came as a surprise to some researchers who were certain that the distinctive pigment signatures of various plastids indicated their origins from a diversity of prokaryotes containing the same pigments (Margulis 1981).

For example, the photosynthetic bacterium *Prochloron* contains chlorophyll a and b, as do the plastids of plants and green algae; cyanobacteria in general contain chlorophyll a and phycobiliproteins, as do the red algae. It seemed logical that ancestors of *Prochloron* must have given rise to green plastids (chloroplasts), and that the cyanobacteria lineage gave rise to red plastids (rhodoplasts). On the of rRNA data, however, this seems not to be the case. The rRNA data thus leave many unsolved questions, such as how did the various pigment signatures evolve and how does it happen that some plastid signatures match those of certain prokaryotes? A multiple origin for plastids is still a distinct possibility, even though pigment "signatures" may not yield the definitive evidence.

Synechococcus has short, thick, rodlike unicells without sheaths. Some species are planktonic in open ocean and are estimated to be responsible for as much as ten percent of ocean primary productivity (Waterbury and Stanier 1981). Some species are highly salt tolerant and others are thermophilic, living in waters as hot as 74°C (Castenholz 1981). *Synechococcus,* like all cyanobacteria, contains chlorophyll a and phycobiliproteins. The electron carriers of photosynthesis are a plant-type ferredoxin, plastoquinone, plastocyanin, and cytochrome f—all similar to those of photosystem I and II in plastids. Water serves as the donor of protons and electrons, and oxygen is generated as a waste product, as with plastids.

The internal membranes of cyanobacteria are arranged in a similar fashion to those of plastids. That is, there are thylakoid membranes in which are embedded the chlorophyll and electron transport molecules. Proton pumping and ATP synthesis in cyanobacteria are also similar to the processes of plastids. The phycobiliproteins are organized in granules called phycobilisomes which stud the outer surface of the thylakoid membranes. This distinctive feature is present in rhodoplasts (plastids of red algae) but not in other plastids (Stanier et al. 1981). Therefore, rhodoplasts may represent one of the more direct and unchanged evolutionary lineages from an original cyanobacterial symbiont. Other plastids, with their own distinctive pigments, must have been subject to considerable genetic mutation and drift as they evolved. Rhodophytes (red algae) also appear first in the fossil record as an identifiable taxon of eukaryotic photosynthesizers. Some of the first photosynthesizers might therefore have resembled the later rhodophytes.

IS *PROCHLORON* A CLOSE RELATIVE OF PLASTIDS?

IN SPITE of the rRNA data, which strongly suggest a cyanobacterial and specifically synechococcal origin for almost all plastids, certain features of another bacterial group (probably cyanobacterial), the prochlorophytes, cannot be ignored. The prochlorophytes are mostly obligate symbionts of invertebrates. They are photosynthetic prokaryotes with chlorophyll a and b, and no phycobilins, just as in chloroplasts (Lewin 1981). However, the carotenoid pigments of one genus *Prochloron* seem to be more similar to those of cyanobacteria than to those of chloroplasts. Photosynthesis is oxygenic in *Prochloron,* although few details are known about the biochemical pathways involved because *Prochloron* has not yet been cultured. The pathways are assumed to involve photosystems I and II, making them similar to those found in plastids.

In trying to understand the evolution of diverse pigments now present in plastids, it should be kept in mind how similar the various chlorophylls really are. Chlorophyll a and b differ slightly in their ring structure and also in their phytyl tails. Furthermore, chlorophyll a is a developmental precursor to b in plastids. Therefore, the evolutionary steps from a plastid with chlorophyll a only to one with chlorophyll b may have been fairly simple. Indeed, the evolution of chlorophyll b from chlorophyll a and the loss of phycobilins seems to have occurred at least three times—and independently—in prochlorophytes (Pallenik and Hazelkorn 1992; Urbach et al. 1992). It could easily have happened in chloroplasts as well. The evolution of diversity in plastids in other algae may have occurred with similar ease. Not only might several distinct acquisitions of plastid symbionts have occurred, but there may have been numerous modifications and secondary losses of pigments to yield the present diversity of photosynthesizers.

Consider, too, that even within a single plant there may be a diversity of plastids, not all of which contribute to photosynthesis. For example, chromoplasts do not make chlorophyll but are rich in carotenoids, giving the reds and yellows of some fruits, flowers, and leaves. Leucoplasts lack all pigments and are sometimes used to store starch.

THE PLASTID ENVELOPE: ARE PLASTIDS POLYPHYLETIC?

TABLE 7.1 organizes the major pieces of evidence for deciphering the evolution of plastid-bearing eukaryotes. First of all, the rRNA sequences of many of these organisms have been analyzed, enabling a family tree to be constructed (e.g., Sogin 1991). The pigment compositions of photosynthesizers confirm and make sense of some of the rRNA-based family tree. This is in spite of the fact that it would probably be quite risky to construct a family tree on the basis of pigment composition alone. Chlorophyll seems to have evolved quickly and easily, suggesting a great tolerance on the part of organisms for variability in pigments.

The number of plastid membranes may be a major clue in understanding plastid diversity. There are some interesting variations in the number of membranes enclosing plastids in various algal groups, a situation unlike that in mitochondria which always have two membranes. The variation may reflect divergent solutions to the problem of extra membranes acquired along with the symbiont, presumably during phagocytosis. In fact, the presence of extra membranes may be evidence for a phagocytotic acquisition. The plastids of plants and of green and red algae (chlorophytes and rhodophytes) are surrounded by two membranes. Plastids of two groups of photosynthesizing protoctists, euglenids and dinoflagellates, are surrounded by three membranes. Plastids of other protoctists—cryptomonads, haptophytes, chrysophytes, xanthophytes, raphidophytes—have four membranes (Lefort-Tran 1983).

TABLE 7.1. Cohesive groups of photosynthetic eukaryotes in approximate order of antiquity (based on rRNA data, pigment composition, and number of plastid membranes).

Antiquity (based upon sequence data)	*Photosynthetic Eukaryote*	*Close Heterotrophic Relative*	*Pigments*	*Number of Plastid Membranes*
▲ Ancient	1. Euglenids	Kinetoplastids, Heterotrophic Euglenids	chl a, b	3
	2. Rhodophytes		chl a, phycobilin	2
▼ Recent	3. Multiple, near-simultaneous branches in *no particular order*			
	a. Dinoflagellates[a]	apicomplexans, ciliates, heterotrophic flagellates	chl a, c	3
	b. Chrysophytes, Xanthophytes, Raphidophytes, Phaeophytes, Bacillariophyta	oomycetes labyrinthulomycota	chl a, c (and in Xanthophytes chl e)	4

c. Glaucocystaphytes *(Cyanophora)*		chl a, phycobilins	2 ?
d. Chlorophytes, Conjugaphytes, Plants		chl a, b	2
e. Cryptophytes	heterotrophic Cryptophytees	chl a, c, phycobilins	4
f. Haptophytes		chl a, c	4

NOTE: Eustigmatophytes with chl a and four plastid membranes do not yet have a place on the tree, nor do Chloroarachnids with chl a and b and 3 or 4 (or 2?) membranes.

[a] And Prymnesiophytes (coccoliths) with chlorophyll a and c and 3 membranes?

An analysis of plastid membranes is done by splitting through the horizontal plane of the lipid bilayer, using freeze-fracture techniques to yield two single layers of lipids. When plasma membranes are split, the two single layers look different depending upon whether they were originally facing in toward the cytoplasm or out toward the environment. The cytoplasmic (protoplasmic) face has a higher density of particles, such as embedded proteins, than does the environmental (exoplasmic) face.

This splitting technique was applied to plastid membranes by Lefort-Tran (1983) with interesting results. When the membranes of two-membrane plastids were fractured, only the inner membrane was oriented in the normal fashion, with the protoplasmic surface facing in and the exoplasmic face on the outside. In contrast, in the second membrane surrounding the plastid, the exoplasmic face appears to be on the inside while the protoplasmic face is on the outside (figure 7.4). The interpretation is that the inner membrane represents the original membrane of the symbiont and that the outer membrane represents the inside-out phagocytotic membrane of the host.

The euglenid plastid has three membranes. The inner membrane seems by its freeze-fracture characteristics to be the original membrane of the symbiont; the second membrane is the inside-out phagocytotic membrane of the original host, and the third membrane is the inside-out phagocytotic membrane of the euglenid itself. Crucial to this interpretation is the idea that euglenids picked up a plastid from another eukaryote, and thus would have picked up a double-membraned structure (figure 7.4). This appears to also be the case for the plastids of dinoflagellates, which are likewise bounded by three membranes.

Plastids surrounded by four membranes are found in several groups of algae, all of which have what is called chlorophyll c, and in one case chlorophyll e. (Chlorophyll c, lacking a phytyl tail, is technically not a chlorophyll but a porphyrin.) Freeze fracture studies suggest that these plastids represent entire photosynthetic eukaryotes that were acquired as symbionts. The cryptophyte algae have single, large plastids bounded by four membranes that show some remnants of the original eukaryotic symbionts (Gibbs 1981; Douglas et al. 1991). Inside the plastid is a double-membraned structure called the nucleomorph, which is presumed to be the remnant of the eukaryotic nucleus of the original symbiont—probably a green alga. There also seem to be remnants of endoplasmic reticulum and

FIGURE 7.4. Membrane topology: two plastid membranes suggest acquisition of a prokaryote; three membranes, the acquisition of a eukaryotic plastid; four membranes, an entire eukaryote with plastid.

ribosomes of eukaryotic size in the space between the second and third membranes. The nucleomorph divides during cell division and seems to contain ribonuclear proteins. Microtubules are apparently not involved in the nucleomorph division. In a relationship analogous to that of the colorless, parasitic plants with the green plants, is *Chilomonas*—a colorless, heterotrophic cryptophyte that has a colorless plastid with a nucleomorph. *Cyathomonas* is also a colorless heterotrophic cryptophyte, but it appears to have never picked up a symbiont (Gibbs 1981).

A POSSIBLE SEQUENCE OF EVENTS IN PLASTID EVOLUTION

ACCORDING TO rRNA analyses two groups of photosynthetic eukaryotes are ancient in comparison to all of the others (table 7.1). These are the euglenids (a group of protoctists, of which *Euglena* is part) and the rhodophytes. The two groups apparently acquired their plastids in independent events.

The euglenids are close relatives of another protoctist lineage: the heterotrophic kinetoplastids, of which the parasitic *Trypanosome* is an example. Euglenids seem to have acquired a double-membraned symbiont, adding to it a phagocytotic membrane to give a total of three membranes surrounding the euglenoplast. Structural evidence thus suggests that euglenids became photosynthetic by picking up a photosynthetic eukaryotic symbiont. It is an intriguing idea, suggesting that some earlier eukaryote, not found on the family tree of extant algae, had already acquired a plastid, thereby making it available for such a union. At some point chlorophyll b must have evolved from chlorophyll a, and phycobilins were lost—an easy feat as evidenced by the lineage of *Prochloron*. Rhodophytes are also of an ancient lineage and appear to have acquired and retained the general pigment characteristics of a cyanobacterial symbiont such as *Synechococcus*.

One conclusion from these observations is that three separate plastid acquisitions occurred in the early eukaryotes:

1. First (with no known surviving examples) was the acquisition of a cyanobacterium with subsequent pigment modifications.
2. Then euglenids picked up that first eukaryotic alga.

3. Finally, rhodophytes acquired another cyanobacterial symbiont and did very little to change its pigments.

What happened subsequently is difficult to interpret. There was a radiation of algae yielding about six photosynthetic groups nearly simultaneously (table 7.1). Four of these groups seem to have acquired eukaryotic symbionts on the basis of their three or four layers of plastid membranes. In these groups chlorophyll a must have evolved into various types of chlorophyll c. In one case, the cryptophytes, phycobilins were retained. In another case, the xanthophytes (yellow algae), chlorophyll e evolved. Might this divergence have evolved from just one acquisition of a eukaryotic symbiont, or are there two, three, or four separate acquisitions here? These observations do not allow a definitive conclusion to be drawn.

Simultaneous with the radiation of the chlorophyll a and c types, two other acquisitions of prokaryotic symbionts are believed to have occurred. Chlorophytes (green algae) acquired and modified the pigments of a bacterium similar to *Synechococcus.* Meanwhile, the ancestor of *Cyanophora,* a member of the glaucocystophyte group, seems to have acquired and retained the pigment signature of another bacterium similar to *Synechococcus. Cyanophora* itself, however, may actually represent a recent event, as its plastid (called a cyanelle) retains the cyanobacterial accessory pigments and a rudimentary cell wall. The cyanelle genome is small (about 177,000 base pairs), but it has not yet lost the capacity to code for a small subunit of ribulose-bisphosphate carboxylase, which most other kinds of plastids have lost to the nucleus.

A reasonable, but still tenuous, conclusion is thus that plastid acquisitions occurred at least five times independently in photosynthetic eukaryotes. Perhaps the ease of acquiring photosynthesizers makes even this a low estimate. Further studies are needed.

It is interesting that the great majority of photosynthetic organisms found in endosymbiotic associations with animals or protoctists today are either cyanobacteria or eukaryotic algae (in turn derived from cyanobacteria). In spite of a great diversity of both chemo- and photo-autotrophy in the bacterial world, only these oxygenic photosynthesizers really caught on as partners with organisms in other kingdoms. One known exception is the ciliate *Kentrophorus,* which associates with purple sulfur bacteria, which presumably photosynthesize both for themselves and for their host (Fauré-Fremiet 1951). Other exceptions are the clams and pogonophoran worms that are found in ocean rift areas; for sustenance they depend entirely on chemoautotrophic bacteria in their tissues, which oxidize sulfide compounds emitted from vents.

PLASTIDS AND SEX

IN THIS chapter the term sex refers to the recombination of DNA from two separate genomes within a single cell. The ability of some plastids to recombine with others of their kind in a population within a cell is evidence of their bacterial ancestry. Plastids replicate their DNA and divide in approximate synchrony with the host cell cycle and also when the host cell is enlarging, such that certain numbers of plastids are usually maintained. Plastids can also fuse, providing an opportunity for genome recombination.

Chlamydomonas, a unicellular green alga (chlorophyte) with a single plastid, is one of the organisms of choice in studies of plastid recombination (Gillham 1978). When *Chlamydomonas* goes through its sexual cycle, in which haploid cells fuse to make a diploid cell, the two plastids also fuse and there is an opportunity for that DNA to recombine. *Chlamydomonas* can survive with a wide range of potentially lethal plastid mutations as long as it is maintained as a facultative heterotroph on nutrient medium. There is, however, one technical problem that has been solved in the laboratory. Under normal circumstances, only one set of plastid genes (usually that of the "+" mating-type parent) ultimately survives in the fused plastid. This means that there would be no opportunity for DNA recombination to occur. By culturing the diploid cells on medium that was lethal to all except biparental plastid genomes, it was found, however, that some cells did maintain two plastids (one from each parent). For example, a biparental plastid containing a gene for erythromycin resistance from one parent and a gene for kanamycin resistance from the other parent would be selected on medium containing both antibiotics. Also it was found that under nonselective conditions about five to ten percent of plastids were biparental. The best method for revealing this phenomenon was to irradiate the "+" parent with ultraviolet light prior to mating, in order to yield a high percentage of biparental plastids (see Gillham 1978). Thus, several distinct kinds of experiments demonstrate the prevalence of recombination events between plastids of differing genomes within *Chlamydomonas*.

Plastid recombination in plants is more difficult to document than it is in algae (table 7.2). One might assume that the hundreds of plastids present in each plant cell would, during fusions, recombine

TABLE 7.2. Mode of plastid inheritance in different species of higher plants. After Gillham 1978.

Biparental Plastid Inheritance	*Maternal Plastid Inheritance*
Ferns	Dicotyledons
Scolopendrium vulgare	*Arabidopsis thaliana*
Gymnosperms	*Arabis albida*
Cryptomeria japonica	*Aubrieta graeca, A. purpurea*
Dicotyledons	*Beta vulgaris*
Acacia decurrens, A. mearnsii	*Capsicum annuum*
Antirrhinum majus *	*Cucurbita maxima*
Borrago officinalis	*Epilobium hirsutum*
Fagopyrum esculentum *	*Gossypium hirsutum*
Geranium bohemicum	*Lactuca sativa*
Hypericum acutum, H. montanum, H. perforatum, H. pulchrum, H. quadrangulum	*Lycopersicum esculentum*
Medicago truncatula	*Mesembryanthemum cordifolium*
Nepeta cataria	*Mimulus quinquevulnerus*
Oenothera (Euoenothera) 28 species	*Mirabilis jalapa*
Oenothera (Raimannia) berteriana	*Nicotiana colossea, N. tabacum*
O. (R.) *odorata*	*Petunia hybrida, P. violacea*
Pelargonium denticulatum, P. filicifolium	*Pharbitis nil*
Pelargonium X Hortorum (zonale)	*Pisum sativum*
Phaseolus vulgaris	*Plantago major*
Rhododendron 11 species	*Primula sinensis, P. vulgaris*
Silene pseudotites	*Stellaria media*
Solanum tuberosum *	*Trifolium pratense*
Monocotyledons	*Viola tricolor*
Chlorophytum comosum, * *C. elatum* *	Monocotyledons
Secale cereale	*Allium cepa, A. fistulosum*
	Avena sativa
	Hordeum vulgare
	Hosta japonica
	Oryza sativa
	Sorghum vulgare
	Triticum aestivum, T. vulgare
	Zea mays

*Plants in which plastid inheritance is predominantly maternal but that show a trace of biparental inheritance.

their DNA, although there would be no definitive evidence if these plastids were all essentially identical. On the other hand, any beneficial mutations might be spread more quickly through a population of plastids in a cell if such fusions and recombinations were taking place.

Despite the experimental difficulties, competition between plastids in a heterogeneous population was studied in one genus of plant: *Oenothera* (the evening primrose). Recombination within the plastid population of *Oenothera* was studied by making interspecific hybrids. *Oenothera* plastids are passed on biparentally. Plastids of *O. hookeri* function normally if transferred into the cytoplasm of either *O. hookeri* or *O. lamarckiana*. However, *O. lamarckiana* plastids become yellow and nonfunctional if transferred into the cytoplasm of *O. hookeri*. Some necessary but unknown factor is apparently not supplied in the cytoplasm of *O. hookeri*. This suggests that after the two species diverged, the plastids of *O. lamarckiana* apparently lost at least one gene still retained by *O. hookeri*.

It appears that some plastids have phenotypic advantages over others in a particular cytoplasm. This is true for *Pelargonium*, another biparental plant genus, known popularly as geraniums. In *Pelargonium* green (functional) plastids are favored over yellow plastids, which are in turn favored over colorless varieties. This is to be expected in a phytosynthesizing organism. Competition between plastids is one of the mechanisms by which plastids manage to maintain useful genes, while lack of competition such as might occur in a colorless, parasitic plant, quickly results in gene loss.

Part of the problem in experimentation with plants is that they are not easily sustained if they carry serious mutations in their plastids. Unlike the situation with *Chlamydomonas* it is difficult or impossible to maintain a plant through a full life cycle as a "facultative heterotroph." Good subjects for such studies are young plants still taking nourishment from the seed, and colorless parasitic plants, which are best able to sustain a serious plastid mutation. Variegated plants, such as coleus, that bear some colorless or partly colored leaves but sufficient green leaves to sustain the plant can also carry potentially lethal mutations. Another problem is that in the sexual cycles of many plants, the plastids are passed on exclusively, or almost exclusively by the maternal partner, probably because of the larger volume of cytoplasm contained in female gametes. Exceptions in which plastids are passed biparentally include *Pelargonium* and *Solanum* (potatoes).

PHOTOSYNTHETIC EUKARYOTES IN THE FOSSIL RECORD

IT IS necessary now to return to the question (first taken up in chapter 1) of the identity of those fossils classified as the first, recognized eukaryotes in the geological record. These fossils include the large coccoids found in rocks 1.7 billion years old in China and the thick filaments, *Grypania,* dated at 2.1 bya in North America. These are not, however, considered to be the first eukaryotes that ever lived. Even as far back as 2.1 billion years ago eukaryotes had probably been evolving for over half a billion years.

A likely identification for these fossilized cells is that of eukaryotic algae, which would already have been quite advanced in having acquired both motility organelles and mitochondria at earlier points in the family tree. But how do we know these fossils to be photosynthesizers? Extant symbioses involving photosynthesizers indicate that such symbioses are often accompanied by an increase in calcium carbonate precipitation as well as increase in cell size. Both conditions appear to pertain to some of these ancient fossils.

THE LARGE CALCIUM CARBONATE PRECIPITATORS: EXTANT SYMBIOSES

THERE SEEMS to be a link between photosynthesis by symbionts and extensive calcification by the host organism. This link is particularly well understood in corals (anthozoan animals) and foraminiferans (granuloreticulosan protoctists, which are like amoebas with shells). Both build calcium carbonate structures (Taylor 1983).

Reef-forming corals have associations with photosynthetic dinoflagellates, a group of motile alga that comprises a large segment of the photosynthetic capacity of plankton. Tropical marine foraminiferans associate with a wide range of algal groups: diatoms, dinoflagellates, chlorophytes, and rhodophytes. Calcium carbonate precipitation is enhanced by photosynthetic activity and stored photosynthate may be important for skeletal growth. Organisms with calcium carbonate structures predominate in the fossil record, with foraminiferans and corals being particularly abundant.

It is interesting to consider the role of photosynthetic symbionts in establishing the diverse lineages of carbonate-accreting organisms. Taylor's experiment (1983) involved Atlantic staghorn coral (*Acropora cervicornis*) and a foraminiferan (*Archais angulatus*). The coral receives about 62 percent of the carbon fixed by the symbionts, and of that about 51 percent is used to form calcium carbonate. The foraminiferan receives about 71 percent of the photosynthate and deposits 63 percent of that in its carbonate housing, called tests. In the dark, these organisms obviously do not photosynthesize, and carbonate deposition decreases dramatically.

Carbonate deposition seems not to be a competing process but a complementary one with photosynthesis. Dissolved carbon dioxide in sea water is in equilibrium with carbonate ions (HCO_3^- and CO_3^{-2}); the former is the preferred carbon source for the photosynthesizer while the latter contributes to the calcium carbonate skeletons. By acting as a sink for CO_2, the photosynthetic symbionts may be enhancing the precipitation of $CaCO_3$ (Taylor 1983).

The importance of photosynthetic symbionts in the fossil record of the foraminiferans has been hypothesized by several authors (e.g., Lee 1989). There are about 35,000 species of foraminiferans recognized in the fossil record. Even today they are so abundant in some marine environments that they seem to "pave" the sea floor, forming "living sands" or "globigerina ooze." The foraminiferans of the tropical and semitropical photic zones have photosynthetic symbionts that seem to make these foramaniferans by far the largest species, visible with the naked eye. An extinct species from the Eocene, *Nummulites* (coin-rock) *gizehensir,* reached 12 cm in diameter. Twenty families of tropical foraminiferans have been evolving larger and larger shells since the Mesozoic, apparently owing to their symbiotic associations. A possible advantage to the symbionts is the fact that the enlarged and translucent foraminiferan shells are like miniature greenhouses, with the equivalent of reinforced frames and thin windows. The shells of the hosts, of course, also protect against some predators (Lee 1989). Phototaxis, directed movement toward light, has been demonstrated in three of the large foraminiferan species, and it is probably present in others, considering that optimal light is at twenty to forty meters in the photic zone.

Reef-building corals not only fossilize well but actually form geographical features, reefs and atolls, so extensive is their skeletal building activity. Reef builders are mainly distinguished from the solitary corals in that they have symbiotic dinoflagellates and live in huge colonies in the photic zone, where the activity of the photosynthesizer enormously enhances calcium carbonate precipitation. The solitary corals, dwelling below the photic zone and without photosynthetic symbionts, precipitate small skeletons and grow much more slowly than the reef builders (e.g., McLaughlin and Zahl 1966).

Thus the presence of especially large, calcium carbonate precipitators such as foraminiferans and corals may be indirect indicators of the presence of symbionts which enhance calcium carbonate precipitation through photosynthetic activity. In fact, a major part of the morphology, physiology, and behavior that are primary identifying characteristics of these organisms, is not so much a function of the animal genes but the genes of their symbionts. That reef-building corals and giant foraminiferans are large, that they precipitate massive skeletons, and that they dwell in the photic zone are all functions of the symbiotic association. By extrapolation, any sudden appearance of large, well-fossilized cells may also be an indication of symbiotic association with photosynthesizers. The large fossil cells, which by their very size made them candidates for the label of earliest recognized eukaryotes, are likely to have been full of photosynthetic symbionts.

THE STORY of the origin of plastids is as well established as that of the mitochondria. And in fact the explanation for why plastids were acquired in the first place is even easier. Symbiotic photosynthesizers provide a valuable service (food production) to their hosts and that is why such associations have been established over and over again in a diversity of groups.

In this chapter we followed the same criteria for ascertaining symbiogenesis that we used in the chapter on mitochondria. Plastids have genomes reduced somewhat from loss of genes and from horizontal transfer to the nucleus. As with the mitochondria this transfer of genes has cemented the relationship between plastids and their hosts. Extreme loss of DNA was described for some parasitic plants and for some protoctist groups. This again brings up the question of how to identify former symbionts that have retained little or no DNA. Our section on plastid metabolism was considerably detailed because the similarity between plastid and cyanobacterial metabolism constitutes a crucial piece of evidence for their relatedness. The complements of pigments in various types of plastids were shown not to be reliable indicators of ancestry, contrary to intuition. However the number of plastid membranes and their orientations do constitute useful information, providing evidence for a polyphyletic origin of plastids. Sexuality in plastids adds an extra level of complexity to the already complex genetics of eukaryotic cells. Populations of both plastids and mitochondria provide many of the essential aspects of the phenotype of a photosynthesizing eukaryote.

VIII

The Controversy About the Origin of Motility Organelles

THE STATUS of mitochondria and plastids as symbionts in the eukaryotic cell is so firmly established that it has attained (albeit only recently) the level of fact in most introductory biology textbooks. The notion that motility organelles in eukaryotic cells might also owe to symbiogenesis is, however, much less certain—indeed, far from certain many would contend. There is, however, a vocal and persistent minority that claims as the preferred hypothesis a symbiotic origin for microtubular systems. The evidence supporting such a symbiotic origin will be the primary topic of this chapter, but interpretations of evidence to the contrary will also be discussed.

STRUCTURE AND FUNCTION

MICROTUBULES ARE tubes thirty nanometers in diameter composed of proteins called tubulins. The cross section of a typical microtubule reveals the underlying substructure, showing thirteen filaments arranged in a ring (figure 8.1); in a few organisms this number may be different, ranging between eleven and fifteen. The shafts of microtubules are composed of two types of tubulin, alpha and beta. Microtubules help maintain the shape of a cell and they enable it to change shape by serving as part of the cytoskeleton. Microtubules also make up the spindles of cell division (meiosis and mitosis).

Microtubules also move the cells, most notably as a specific structure called a motility organelle. Motility organelles are complex structures with a distinctive "9 + 2" arrangement of microtubules (figure 8.1). Occurring with the microtubules are numerous microtubule associated proteins (MAPs). One of the most notable proteins, axonemal dynein, serves as an ATP-based motor causing the sliding of microtubules, which in turn bends the motility organelle that moves the organism. Motility organelles (cilia) were first observed by A. van Leeuwenhoek (1623–1723) on rotifers. "I paid

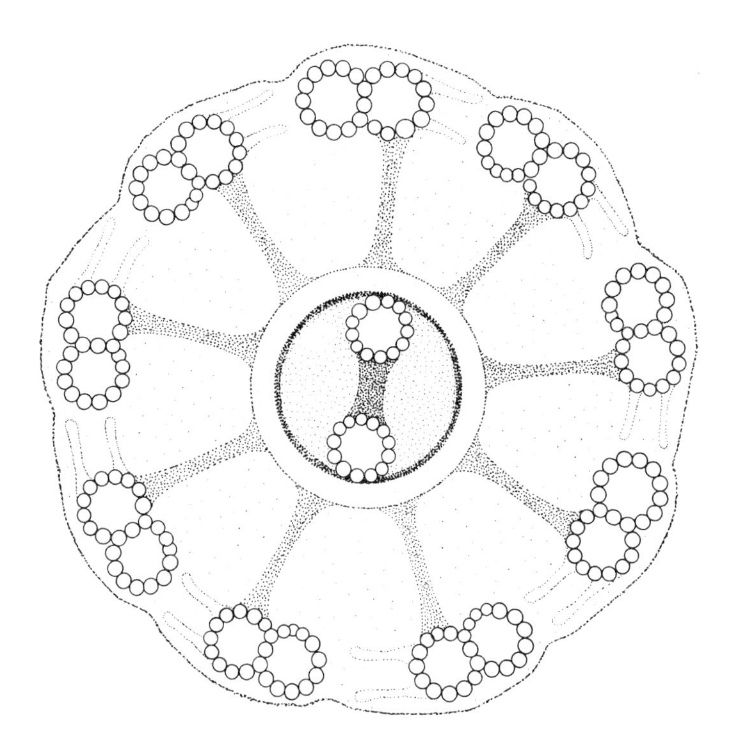

FIGURE 8.1. Microtubules in a 9 + 2 configuration. Drawing by C. Nichols.

great attention to their revolving toothed wheel work; and I saw that an incredibly great motion was brought about by said instrument in the water around it" (Dobell 1988).

The controversy about the evolutionary origin of microtubules extends even to the terminology. The motility organelles of eukaryotes are so distinctive in their structure and so different from the flagella of bacteria that the term flagellum is inappropriate for the eukaryote organelle. The use of the term flagellum for both the prokaryotic and eukaryotic motility organelles predates the elucidation of the fundamental biochemical and molecular differences between the two structures. As to what word should replace flagellum when used for eukaryotes, a number of suggestions have been made. The terms cilia and sperm tails have only limited use (e.g., for ciliates and sperm). "Undulipodium" has been recommended and so has "9 + 2" and "motility organelle." Avoiding further discussion of semantics and etymology, we will use the general term "motility organelle" in this book, as we did in a previous book (1986) on eukaryotic cells.

THE MICROTUBULE ORGANIZING CENTER

MICROTUBULE GROWTH is initiated at sites called microtubule organizing centers: MTOCs. These are typically found at the base (or "minus" end) of a microtubular structure, such as a motility organelle or spindles in cell division. MTOCs appear to be highly conserved, in that they apparently are universal in eukaryotes. Specific microtubule structures, on the other hand, seem to be highly divergent and variable. It is the highly conserved nature of MTOCs as well as their unique replication behavior, described in the next section, that suggests that the microtubules themselves might be highly integrated symbionts acquired by the host cell.

A microtubule organizing center (MTOC) is a specific area of the cell that nucleates the growth of microtubules. It usually is found at the base of an individual motility organelle or at the origin of the microtubules forming spindle fibers during cell division. The term MTOC is often used interchangeably with "basal body" and "centrosome." All are nearly identical structures, showing a "9 + 0" pattern of microtubules (figure 8.2). This "9 + 0" pattern has been described as "a squirrel cage" by Randall and Hopkins (1963).

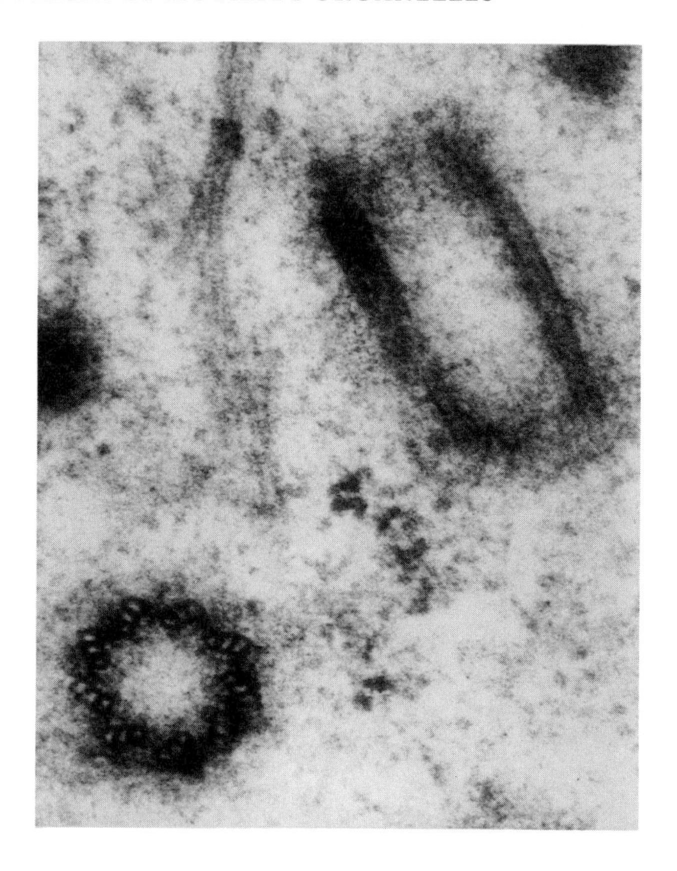

FIGURE 8.2. Microtubules in a 9 + 0 configuration. Reprinted from Margulis 1993. Photo by E. R. Dirksen.

For those many organisms in which the MTOC is always associated with a 9 + 0 structure, the term MTOC is routinely used as a virtual synonym for the 9 + 0 microtubule structure itself. There are many organisms, however, that maintain MTOCs with no associated microtubule structures at all. Plants are an example; they have no centriolar structure, but do have an MTOC "area" in which spindles are initiated. The MTOC, therefore, is actually something more elusive (less structurally visible) than the squirrel cage with which it is often associated.

Sluder (1989), working with the centrosomes of echinoderms, has done much to elucidate the notion of the MTOC. The two centrosomes of animal cells consist of a pair of 9 + 0 structures arranged at right angles to each other, surrounded by fibrous "pericentriolar material." It is this difficult-to-isolate fibrous material which appears to be the actual organizing center. The functions of centrosomes are to nucleate the microtubules of spindles used in cell division and to double or reproduce just before cell division (ensuring that each daughter cell will receive two centrosomes). Centrosomes are able to double without nuclear participation, as shown in cells with their nuclei removed (enucleated cells). Furthermore, echinoderm centrosomes may have the centriolar (microtubule) structures (9 + 0) experimentally removed, in which case the MTOC (apparently the fibrous pericentriolar material) continues to double appropriately in the cell cycle (Sluder, Miller and Rieder 1989).

As many previous researchers have pointed out, the phenomenon of the replicating MTOC could be explained more easily if it contained DNA or RNA—that is, a genome. Therefore, numerous researchers have searched for nucleic acids in MTOCs (and in the associated 9 + 0 structures, which are more easily isolated). Most of these researchers have discovered that MTOCs are uncooperative little entities, for the search has been fraught with the difficulty of getting reliable preparations. A new method for reconstituting centriole duplication in biochemical extracts from clam oocytes is now available (Palazzo 1992), and this may pave the way toward an understanding of the genetics of the MTOC components.

DO MICROTUBULE ORGANIZING CENTERS HAVE DNA OR RNA?

MICROTUBULE ORGANIZING centers (MTOCs) have been the subjects of a long search for the presence of DNA in or around them. MTOCs have been observed to replicate with the same apparent autonomy as mitochondria and plastids display. It seemed to make sense therefore to apply the same techniques to MTOCs that were being used to identify DNA in plastids and mitochondria. However, those techniques have not yielded confirmation of DNA in MTOCs.

The search for DNA in MTOCs proceeded approximately as follows. (The reader may wish to consult the summary of these events in table 8.1.)

Tetrahymena pyriformis was fractionated by Hoffman (1965), as

TABLE 8.1. Search for DNA or RNA in MTOCs

	Orcinol Assay for RNA	*Diphenylamine Assay for DNA*	*Tritiated Thymidine Uptake*	*Acridine Orange Stain for DNA, RNA*	*DNAse*	*RNAse*	*DNA Synthesis Inhibitors*	*DNA Probes*	*DNA Antibody*	*Genetics*
Hoffman 1965	+(RNA)	+(DNA)								
Randall and Disbrey 1965			+(DNA)	+(RNA,DNA)	+(DNA)	+(RNA)				
Younger et al. 1965			+(DNA)				—			
Heidemann & Kirschner 1975					—	+(RNA)				
Dippell 1976						+(RNA)				
Hartmann et al. 1977				+(RNA)	—	+(RNA)				

Hall et al. 1989	+(DNA)		
Johnson & Rosenbaum 1990		—	
Johnson & Dutcher 1991			—

NOTE: *All* of these results are either equivocal or controversial in some way. For further details, see the accompanying text. Isolating *pure* MTOCs consistently has been the problem.
+ = indicates positive results.
− = a negative result.

well as by previous investigators (Seaman 1959), in order to collect MTOCs at the bases of the motility organelles of this ciliate. The preparation was probably not free of contaminating nuclear DNA and RNA; so the conclusion that 1.2 to 2.6 percent RNA was present (on the basis of the orcinol assay) and that 0 to 1.0 percent DNA was present (on the basis of the diphenylamine assay) was equivocal. The DNA levels were near the limits of detection for the method.

In the same year Randall and Disbrey (1965) labeled *Tetrahymena* with tritiated (radioactive) thymidine, which should be taken up into a DNA strand just as would nonradioactive thymidine—thereby becoming a convenient indicator. They then isolated the MTOC-rich cortex and examined it using an autoradiography technique (to detect the radioactive molecule). The cortex was simultaneously stained with acridine orange. Both tests indicated the presence of DNA in very small amounts. A weight of 2×10^{-16} gm was a rough estimate. A temperature shock treatment had been used to get the culture of organisms to divide synchronously. Not until two hours after shock treatment was DNA detectable in the MTOCs. This suggested that nucleic acid was detectable only when the MTOCs were about to divide. The color of the acridine orange stain (which fluoresces yellow-green in the presence of DNA) disappeared after treatment of the MTOCs with DNAse. This left particles that fluoresced orange-red (indicating RNA), and that color was extinguished with RNAse. Randall and Disbrey wondered, as have many subsequent observers, whether DNA from a disrupted nucleus might be attached to MTOCs. Interestingly, it was found that MTOCs treated with DNAse and then recontaminated (or contaminated for the first time?) did not show evidence of DNA.

Younger et al. (1972) looked for evidence of DNA synthesis during the rapid production of oral membranelles (the cilia around the mouth) and thousands of new MTOCs by the ciliate *Stentor*. (This organism can be made to spontaneously lose its membranelles by placing it in a salt solution.) Specifically, Younger and colleagues tested for incorporation of tritiated thymidine and for sensitivity to DNA synthesis inhibitors. DNA synthesis apparently did not accompany membranelle synthesis, although DNA could be detected in the MTOCs by autoradiography. Mitochondrial contamination was considered to be a plausible interpretation for the mixed results. Indeed, mitochondria are often found closely associated with MTOCs, especially those at the base of motility organelles. Difficulties in isolating "clean" MTOCs continues to be a barrier to getting unequivocal results in these sorts of experiments.

Hartman et al. (1974) developed a new procedure to isolate MTOCs from the cortices of *Tetrahymena*—a procedure that they believed would yield MTOCs free of contaminating DNA and RNA. Their preparations showed a yellow-green color when stained with acridine orange, an indication of DNA. However, they were able to extinguish the color with RNAse and not DNAse. They concluded that there was single-stranded RNA in the MTOCs, so tightly configured that it yielded a DNA-like (double-stranded) color. This stain gave a positive reaction, regardless of the timing in the cell cycle.

Heidemann and Kirschner (1975) were able to induce formation of asters (starlike structures composed of an MTOC with numerous microtubules emanating from it) in frog eggs by using the purified MTOCs of either *Chlamydomonas* or *Tetrahymena.* This illustrated the universality of the MTOC not only because of the interspecific interaction but because the particular MTOCs were functioning at the bases of motility organelles in the protists and also acting as spindle organizers in the frog. Heidemann et al. (1977) elaborated on this study by attempting to alter MTOC function with DNAase, proteolytic enzymes, and RNAase. They found that RNase altered the function, as did proteolytic enzymes, probably by causing structural damage to the 9 + 0 structure. Heidemann and collaborators showed that RNA in MTOCs was necessary for initiation of microtubules but not for their elongation. They also concluded that the quantity of RNA was about 5×10^{-16} gm per basal body. They noted that a "surrounding cloud" around the MTOC 9 + 0 structure seemed to be responsible for microtubule initiation; this cloud has since come to be known as the "pericentriolar schmutz." This observation was interesting in light of the numerous MTOCs that function without the distinctive 9 + 0 structure—a point that will be further discussed in another section.

Meanwhile, Dippell (1976) found that both pronase (an enzyme that breaks down protein) and RNAse disrupted the fine structure of the *Paramecium* cortex. The cortex is an unusual semiautonomous structure, composed of microtubules, and it is the topic of a later section.

THE "UNI" GENES OF *CHLAMYDOMONAS*

SOME OF the most interesting but controversial work of late has involved a group of linked genes in *Chlamydomonas* called the "uni linkage group." Mutant *Chlamydomonas,* bearing only one motility organelle, the one furthest from the eye spot, were found to be carrying a defective gene "uni" that was not linked to known nuclear genes. The defective gene was however linked to four other mutant loci that also affected the assembly of motility organelles. The problem with uni seemed to be with the structure of the MTOC (Huang et al. 1982).

Uni was subsequently found to be linked to eleven other loci, and the entire uni linkage group was defined through classical genetics as a circular chromosome of about a hundred "map units."[1] The linked genes and their phenotypes are as follows:

uni 1 no motility organelles near eye spot
sun 1 modifies uni 1 by suppressing it
enh 1 modifies uni 1 by enhancing it
fla 9 no motility organelles at 32°C; rapid production at 21°; slow resorption
fla 10 no motility organelles at 32°; rapid production at 21°; rapid resorption
fla 11 no motility organelles at 32°; slow production at 21°; slow resorption
fla 12 no motility organelles at 32°; slow production at 21°; slow resorption
fla 13 no motility organelles at 32°; slow production at 21°; slow resorption
pf 7 short stubs
pf 8 no motility organelles
pf 10 abnormal motility (irregular circles)
pf 29 abnormal motility (jerky linear path)

The linked genes all seem to be inherited in Mendelian fashion—that is, as though they were nuclear genes. Ramanis and Luck (1986)

[1] Map units are relative distances on a chromosome. Positions on a chromosome that cross over frequently with other positions are further away in map units, which are thus representative of crossover frequency. Closer positions have fewer map units between them (lower crossover frequency).

noted that five days before meiosis in *Chlamydomonas* the frequency of recombination increased in the uni linkage group if the temperature was increased from 17 to 32°C. It seems that replication and recombination may be occurring at this time, in the uni group.

An additional gene found in this linkage group, apm 1, confers resistance to the herbicide amiprophos-methyl and to close analogues. Another gene (apm 2) with similar properties was found on a nuclear chromosome. The gene products of apm 1 and 2 somehow interact to confer resistance (James et al. 1988).

The uni linkage group was fragmented and partially cloned by Hall et al. (1989) and the clones were used as probes to localize the linkage group in the cell, as well as to determine size and form of the DNA. By molecular techniques, the uni linkage group was found to be a linear molecule of six to nine megabases (a megabase is one million bases) that is located in or on the MTOCs. Here, at last, seemed to be the direct evidence that proponents of a symbiotic origin for motility organelles had been seeking. Here was a large genome, carrying at least thirteen genes—twelve of them related to motility organelle functions—and the uni DNA probe selectively lit up the MTOCs. However, these data became the center of a controversy about interpretation, one that is not yet resolved.

The fact that this apparent genome of the MTOCs segregated in a Mendelian fashion could be explained by the presence of two MTOCs (acting as centrioles) per cell which help the nuclear chromosomes to separate and, in doing so, segregate themselves. The quantity of DNA found by Hall et al. would be hard-pressed to fit into a space of the lumen of the basal body, but other areas of the structure could be involved. The conclusion that the chromosome was linear and not circular as the classical maps had indicated could be explained in at least two ways. First, the chromosome may alternate between a circular and a linear form. Second, the genome may replicate by a mechanism like that of a bacterial virus. That is, a linear sequence of genes may be represented as abc or bca or cab. This would appear to be a circular map, even though the chromosome might well be linear:

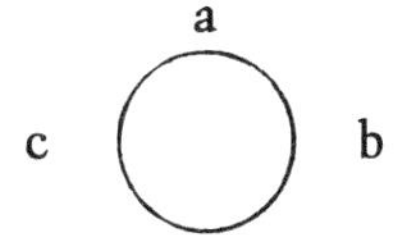

This publication of Hall et al. (1989) received a great deal of attention and it was quickly followed by two reports refuting it.

Johnson and Rosenbaum (1990) examined whether a nonspecific anti-DNA antibody would attach to MTOCs, as well as all other DNA in the *Chlamydomonas* cell, including the nucleus, mitochondria, and plastids. The results were disappointing (at least to proponents of a symbiotic theory). MTOCs did not react at all with the antibody, in spite of the large amount of DNA (6–9 megabases) purported to be there. *Chlamydomonas* plastids with 8–10 nucleoids (DNA-rich areas), each containing 2 megabases of DNA, stained well with the nonspecific probe. Mitochondria contain approximately 15–150 kb (kilobases) per mitochondrion, depending on how many copies of the genome there are and how many mitochondria. The mitochondria picked up very little of the probe, however they have considerably less DNA than do plastids.

Johnson and Rosenbaum claimed that their method was less disruptive to the cell than that of Hall et al. and therefore less likely to risk contamination of the MTOCs by nuclear DNA. Such contamination of the MTOCs might yield a false positive result for presence of DNA. This is what Johnson and Rosenbaum seem to suggest might have happened to the preparation of Hall et al. However, if the MTOCs of Hall et al. were contaminated, it would be an extremely specific sort of contamination. Why was the putative contaminant DNA so uniquely attracted to the MTOCs of the disrupted cells? Furthermore, is it correct to assume that Johnson and Rosenbaum's antibody probe would react with MTOC DNA? After all, if MTOC DNA exists, it may be condensed differently from other DNAs, such as the control DNAs that Johnson and Rosenbaum relied upon. Unfortunately a positive control is not possible for this experiment, which relies on a *negative* result to contradict the earlier finding.

Johnson and Dutcher (1991) followed with a paper supplying more indirect evidence that the uni linkage group is not localized in the MTOCs of *Chlamydomonas.* These authors argued that the fact that the uni group is present in single copy in haploid cells (which have two centrosomes) and in double copy in diploid cells means that it is behaving no differently from any nuclear gene and probably is nuclear. Furthermore, a mutant *Chlamydomonas* without visible MTOC structure present continued to show the presence of the uni linkage group. The simplest hypothesis to fit these data, say Johnson and Dutcher, is that the uni genes are nuclear.

There is an alternative hypothesis suggested by Johnson and Dutcher: that in haploid cells the genome localizes to only one of the MTOCs, and that in cells lacking MTOCs the linkage group exists elsewhere. Although the authors quickly dismiss this hypothesis, we ourselves find it intriguing, and it will be developed a little further in the next section because it is consistent with some other observations about MTOCs.

AN ATTEMPT TO RESOLVE THE CONTROVERSY

FOR THE sake of discussion, let us assume that MTOCs do contain nucleic acid genomes and that the ability of an MTOC to replicate is dependent somehow upon the presence of the genome. During each mitotic cell division of a diploid cell, two centrosomes (each containing two MTOCs, which in turn each contain a genome) replicate so that each daughter cell will receive a pair of centrosomes. Haploid cells, having undergone meiotic divisions, also have two centrosomes but only one is passed on to the diploid zygote in fertilization. Transmittal of only a single centrosome by a gamete is important because a zygote receiving four centrosomes might proceed into irregular mitotic divisions.

There is an additional problem, however, pointed out by Sluder et al. (1989). In the animal zygotes they studied, the centrosomes go through a division before the first zygotic division. This means that there is a potential for a zygote to develop four centrosomes with four MTOCs, which could result in irregular mitosis. This does not occur, apparently because the maternal centrosome (in the case of starfish) loses its ability to replicate and is eventually lost. The maternal centrosome also loses its ability to organize microtubules. This maternal centrosome was found to consist of one (not two) MTOCs.

Sluder et al. (1989) hypothesized that whatever enables the maternal centrosomes to replicate and function must have been lost. We suggest that this eliminated feature might be the MTOC genome, or DNA. In other animals, such as mammals, there may be a similar mechanism to ensure the requisite paring down of centrosomes for the zygote, but in that case it would be the paternal centrosomes that do not replicate. *Chlamydomonas* may utilize a similar approach during its sexual cycle to retain only two centrosomes. The fact that haploid *Chlamydomonas* have one uni group and diploids have two uni groups would be entirely consistent with a mechanism for retaining exactly two functional centrosomes in a zygote. In fact, Johnson and Dutcher's (1991) work would seem to supply a mechanism for Sluder's observation, albeit in a different species (figure 8.3).

As to Johnson and Dutcher's alternative hypothesis that *Chlamydomonas* mutants lacking MTOC microtubule structures have a uni group localized elsewhere, it should be pointed out that the structural aspect (9 + 0 microtubule arrangement) of MTOCs has long been known to be variable. Plants, for example, organize their microtubules during cell division with dense plaquelike areas that are considered to be MTOCs, but which have no trace of a 9 + 0 microtubular structure. The 9 + 0 structure is a conveniently visible marker for most but not all MTOCs.

It may seem as though the rules are being changed in the middle of the game, that 9 + 0 structures are markers for MTOCs, but a lack of such structures does not signify a lack of MTOCs. The MTOC has thus turned out to be a complicated system, and the data presented here show few signs of converging easily toward one particular conclusion as to how MTOCs function and whether or not different types of MTOCs might be involved in different roles.

A NONSYMBIOTIC ORIGIN FOR MOTILITY ORGANELLES?

SURPRISINGLY FEW detailed hypotheses have been proposed to explain how motility organelles evolved. The proposal of Cavalier-Smith (1975) is the only one we are aware of that attempts to consider the data and observations about microtubules and motility organelles and to incorporate them in a nonsymbiotic model.

Cavalier-Smith (1975) noted that a weak link in the entire symbiotic theory for cell evolution is the origin of the various components of the host cytoplasm, such as the nucleus, endoplasmic reticulum, and cytoskeleton. Endocytosis (such as phagocytosis) has been claimed by the symbiotic theorists as an important mechanism by which a pre-eukaryotic host cell might have acquired symbionts. Therefore, one of the keys to finding a prokaryotic descendent of that original host is to find a unique prokaryote with some aspect of a phagocytotic system. Cavalier-Smith argues (and we think correctly so) that the very same phagocytotic system, involving flexible,

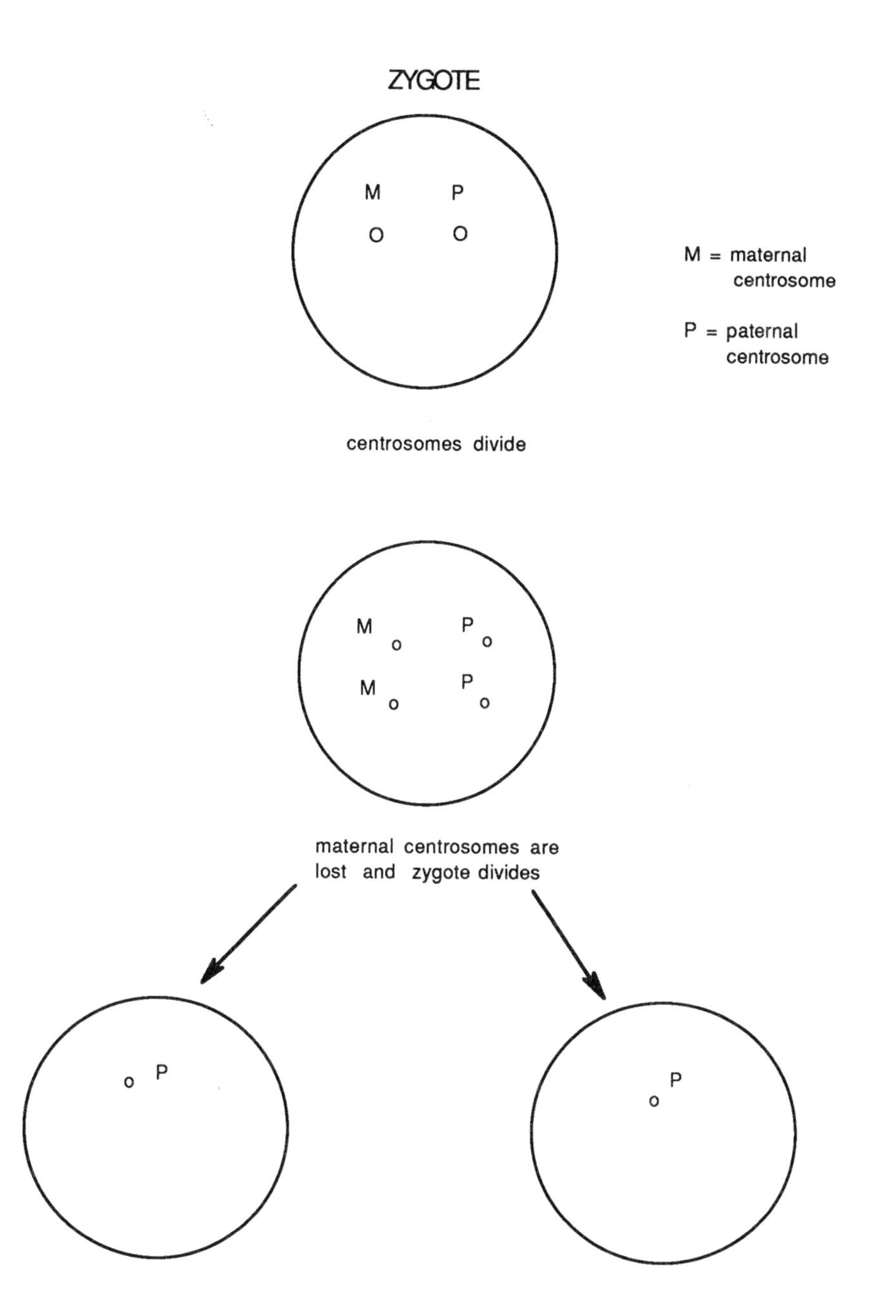

FIGURE 8.3. The apparent haploid-diploid cycle of the uni linkage group fits Sluder's observations on centriole replication.

wall-less membranes, could also account for the various internal membranes of eukaryotes. These would be derived from pinched-off invaginations of the outer membrane.

Such a mechanism may very well account for many of the internal membrane systems such as endoplasmic reticulum, nuclear membrane, and golgi—an idea we discussed in chapter 4. As to an origin for mitochondria and plastids as invaginated compartments, even Cavalier-Smith (1975) no longer considers this a viable hypothesis. The evidence is so overwhelming that these two organelles are derived from symbiotic prokaryotes that an alternative is not even presented in this volume. The motility organelle story, however, is far more ambiguous.

Cavalier-Smith's alternative to the symbiotic hypothesis begins with the contention that microtubule spindles used to move chromosomes in cell division would have been a necessary aspect of cell division in the pre-eukaryotic or early eukaryotic cell. The mobility and flexibility of the cell membrane might interfere with chromosome segregation (which in many prokaryotes depends upon attachment of chromosomes to a stable membrane). Microtubules (as spindles) are nearly universal in eukaryotes and therefore must have arrived as an intact system in that pre-eukaryotic or early eukaryotic cell. Mitotic-like division (with microtubules) would have been a necessary mechanism, owing to the challenge of flexible eukaryotic membranes. Likewise, sexual fusion, recombination, and meiosis would also have evolved in this pre- or early eukaryote. The flexibility of the cell membrane would have enabled fusion between cells and this in turn would have necessitated a mechanism (meiosis) for preventing the genome and organism from growing larger and larger from repeated fusions. After this evolution of mitosis and meiosis, according to Cavalier-Smith, the microtubules of the spindle would have been transferred out to the periphery of the cell and inserted into an out-pocketing of the membrane, thereby becoming the motility organelles. Other microtubular functions in the cell, such as components of the cytoskeleton, were also derived from these original spindle microtubules and would have been necessary in larger wall-less cells.

It seems to us that this kind of scenario for the origin of mitotic division and microtubules in a prokaryotic ancestor is an argument from the viewpoint of incredulity. That is, because the scenario's proponents find it unbelievable that a prokaryote could divide in a wall-less, flexible-membraned state, then it must be that prokaryotes that did evolve into this state also had some special mechanism like mitosis (and all of its associated structures). There are in fact many large, flexible, wall-less (or nearly wall-less) prokaryotes in the Archaebacteria—such as *Thermoplasma* and *Sulfolobus*—that apparently do get along without mitotic-like division (chapter 4). Obviously, the search for a tubulin-like protein in a prokaryote is crucial for deciding the question. If found in an archaebacterium, then tubulin might well have evolved in eukaryotes as Cavalier-Smith suggests. If spirochetes or some other eubacteria have it, this supports a symbiotic origin. The question awaits more data. But a preliminary piece of evidence in favor of the hypothesis that tubulin arrived with the host archaebacterium was developed by Sioud et al. (1987). They determined that *Halobacterium* (a branch of Archaebacteria) reacted positively with anti-tubulin antibodies and that growth was inhibited by vincristine and nocodazole (tubulin inhibitors).

OTHER ARGUMENTS AGAINST A SYMBIOTIC ORIGIN

OTHER AUTHORS have pointed out, correctly, that some MTOCs do not appear to have any particular continuity at all, if one follows their progress in the cell cycle by looking at the associated 9 + 0 microtubule structure. Lack of continuity, and de novo synthesis of new MTOCs (or essential structures) would of course suggest that there is no reason at all to look for DNA or RNA or to suspect that the MTOCs are autonomous. However, there is a problem with following only the 9 + 0 structures, when it is apparent that the diffuse and elusive pericentriolar material is the actual MTOC.

Fulton and Dingle (1971) discovered that 9 + 0 microtubule structures arose de novo in the protoctist *Naeglaria,* which transforms from an amoeboid form (lacking 9 + 0 structures) to a form with motility organelles. No precursor for the 9 + 0 structures was observed. Unfortunately, it is the nature of the pericentriolar material that it could well elude the observer. After a careful and methodical search, Fulton and Dingle found no evidence for a definitive pericentriolar area. It has, however, been noted by several researchers (e.g., Wheatley 1982) that there is a large range in pericentriolar areas in respect to size, density, location, and distinction in the cell.

A definite conclusion that a pericentriolar area is entirely absent may not be possible.

Pickett-Heaps (1971) cites examples in which MTOCs appear to form near the nuclear envelope in several algal species. He confirms that in numerous species MTOCs exist and function without any type of associated 9 + 0 structure. The question as to whether or not such an ambiguous entity appears in certain algae de novo, is not answered, for the reasons stated above concerning Fulton and Dingle's work. Pickett-Heaps is reluctant, however, to assume that the diffuse pericentriolar material constitutes the actual replicating, autonomous structures that the symbiotic theorists claim it does. It would be equally reasonable to suppose that any other cell organelle of uncertain origin could have a similar autonomy (and possible deviation from undetectable precursors).

Whether or not "autonomy" is a clearly defined trait for any organelle, and just how the pericentriolar material is to be interpreted, remains an outstanding question, only partially answered by some of the evidence presented in this section. It bears repeating here that the various researchers pursuing a symbiotic origin for hydrogenosomes and microbodies (peroxisomes), *neither of which contain DNA,* encounter many of these same problems. In the next section an alternative hypothesis will be developed that motility organelles were of symbiotic origin.

WERE MOTILITY ORGANELLES SYMBIOTIC BACTERIA?

THE NOTION that motility organelles are symbiotic bacteria is a minority viewpoint. Because this minority viewpoint is seldom summarized in the literature, we will discuss in the next few sections those observations and experimental results that are consistent with a bacterial origin of motility organelles and consistent with what may be a DNA genome in the MTOC. We do so not with any assertion that evidence in favor of symbiosis is conclusive, but to indicate that this intriguing problem is alive and well and that the symbiotic hypothesis for motility organelles has in fact not been falsified by direct, positive experimental results.

We begin by examining evidence that the cortex in ciliates may possess an independent genetics. The cortex of some ciliates (a complex outer covering rich in MTOCs and cilia) displays a rather unusual autonomy, especially when repairing or adjusting to wounds and other mishaps. Several researchers have perfected microsurgical techniques with which to introduce specific wounds and grafts to the cortices of ciliates and to follow the repair or adjustment process.

Some specific manipulations to the cortex have been called cortical "mutations" in that they may be passed down generation after generation undiluted or unaltered in asexually reproducing ciliates. A replication of the MTOCs and associated structures of the altered areas is involved; however there is no change in the nuclear DNA. The mutation, if it is correct to call it that, is non-nuclear and may well be involved with the replicating system of the MTOCs.

Two types of cortical mutations in *Paramecium* may be caused by a failure of a mating pair to separate quickly or at all after exchanging DNA (in the form of haploid micronuclei). The pair join in anterior to posterior position, and if they separate only slowly a little piece of the cortex of one might remain grafted upside down to the cortex of the other (figure 8.4). Some of these mutants have been called "twisty" because the patch of upside down cortex causes an irregular swimming pattern. This is because the MTOCs and all of their associated microtubules form an asymmetric complex, a piece of upside-down cortex that may be detected. The grafted piece replicates with every cell division, along with all of the other cortical structures, in what may be typical MTOC fashion. What is interesting is that the upside-down orientation is also maintained generation after generation through hundreds of cell divisions. One would expect, if the cortical graft (mutation) were under nuclear control, that subsequent MTOC units would be formed and would be arranged in correct orientation (e.g., Sonneborn 1970).

Similarly, if two mating paramecia fail to separate at all, they become "mutant" doublets, and this trait too is passed down through hundreds of generations. Introduction of new nuclear DNA, via a sexual cycle, also has no influence on either of these mutations, twisty or doublet (Sonneborn 1963).

It has been observed that populations of other ciliates, such as *Tetrahymena* and *Euplotes,* reproduce true to "corticotype" regardless of what nuclear DNA they might be carrying. That is, particular patterns of cortical units are reproduced generation after generation in what looks like an autonomous mechanism (Nanney 1968; Hufnagel and Torch 1967).

A mirror-image doublet of *Pleurotricha,* created by microsurgery

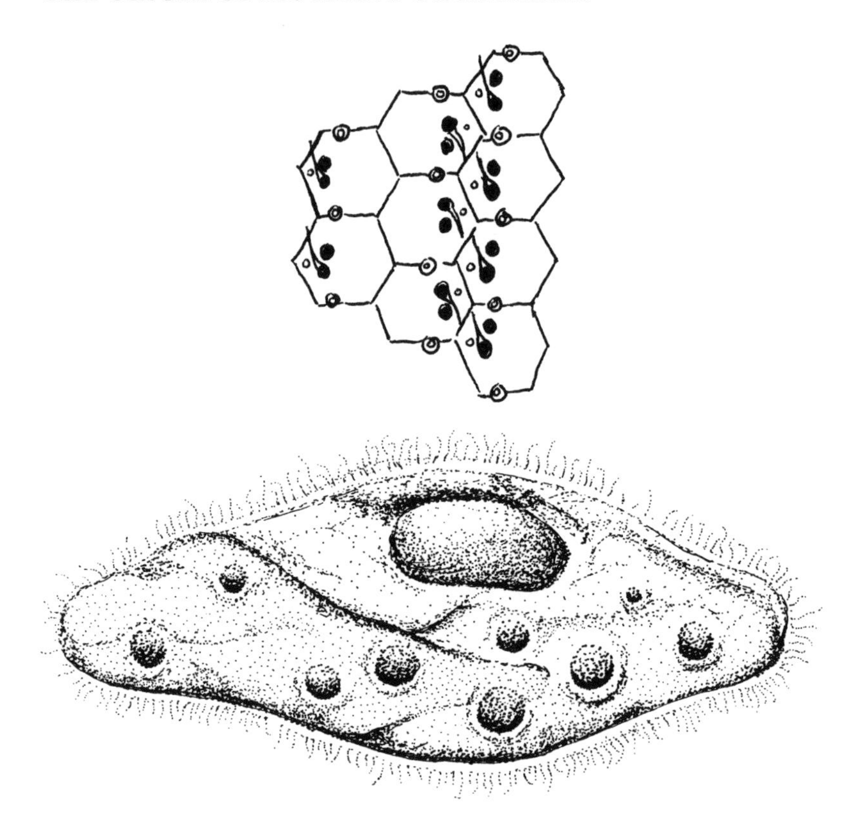

FIGURE 8.4. A: Orientation in the ciliate cortex; note the asymmetry of the various structures in each hexagon; this makes it possible to see that the middle row is upside down.

B: This shows an entire ciliate; each of its many cilia are based in one of the hexagon-like units. Drawings by C. Nichols.

by Grimes et al. (1980), showed the maintenance of mirror-image pattern of cortical structures in the mirror-image ciliate and its descendants. However in this case, the individual structures constituting the whole pattern "reverted" to the former arrangement, no longer mirror image. This suggests that in *Pleurotricha* it is the overall pattern that may be considered to be autonomous from nuclear control.

Grimes (1976) destroyed with a laser microbeam the primordial areas that give rise to cirri (bundles of cilia) in *Oxytricha.* This resulted in an inability of those areas to produce more cirri. The microbeam presumably destroyed the replicating areas in that region of the cortex. It was observed that *Oxytricha* goes through an encystment stage, during which all traces of microtubules are lost. Upon encystment, microtubules are resynthesized in the original positions and orientations. Some cortical mutants of *Oxytricha* have extra rows of cirri and these extra rows are lost and then faithfully regained in the encystment-reemergence process. However, some extra rows are permanently lost, a difficult result to interpret. It suggests that for at least some microtubule structures of *Oxytricha,* the information for positioning some microtubules remains through encystment (but not in the nucleus, as was also demonstrated). For other microtubules the information is lost or altered and a full repair of the mutation is possible (Hammersmith and Grimes 1981).

Could all of these observations possibly be explained by noting that some ciliates are incapable of getting and retaining cortical mutations? Cortical mutation appears to be a tendency primarily in a subphylum of ciliates, Cyrtophora, which includes *Tetrahymena, Euplotes,* and *Paramecium*; members of this group may have in common a slowness or even lack of repair by nuclear coding of any damage to the cortex. The two ciliates that yielded more ambiguous (albeit interesting) results, *Pleurotricha* and *Oxytricha,* are in the subphylum Postciliodematophora—a group that also includes *Stentor,* a ciliate known to readily rearrange and repair cortical damage. Perhaps the ambiguous results from ciliates of this subphylum are due to a partial tendency of the nucleus to override cortical information and render some repairs in *Pleurotricha* and *Oxytricha.*

The various authors of papers on cortical mutations have generally been cautious in proposing mechanisms for the phenomena observed. Nevertheless, there does seem to be general agreement that some sort of information separate from the nuclear DNA controls the orientation and position of microtubular structures in some ciliates. This, we believe, is consistent with the nature of the mutations of the uni linkage group, possibly part of the genome of the *Chlamydomonas* MTOCs. That is, many of the genes of the uni linkage group are involved with the orientation and positioning of microtubules. If with every cell division each MTOC of the ciliate cortex replicates itself, along with a small genome, this would provide a mechanism for the faithful replication and passage of cortical mutations.

If, for example, a product of an MTOC gene is some type of factor that is produced in a concentration gradient from one side of the

MTOC, thus directing the orientation of microtubule growth, then a change in position of the MTOC or a mutation could change the orientation of the gradient and the subsequent orientation of the microtubules. Thus a manipulation that appears to be strictly nongenetic, a graft, might actually be indirectly genetic and involve the DNA of the MTOCs. (We have based this argument on ideas proposed by Frankel 1974 and Grimes 1982.)

There are two possibilities for the apparent limitation of cortical mutations to certain groups of ciliates. One is that as these ciliate groups evolved from the earliest protoctist groups they retained an unusual genetic autonomy in their MTOCs—a power to essentially override the ability of the nucleus to repair damage to the cortex. Other protoctists and other ciliate groups may have lost this autonomy, and thus damage to the MTOCs would be repaired by molecules coded by the nucleus. The fact that mutations of microtubule orientation have been found in a very different organisms, ciliates and *Chlamydomonas* (green algae) for example, suggests a second possibility. That is, the ciliates are peculiar in having all of their MTOCs arranged on the surface, easy to stain and observe, and these MTOCs are fixed in regular asymmetrical patterns so that changes in orientation can be readily seen. It may well be that other eukaryotes also get the equivalent of cortical mutations (mutations of MTOC DNA) but that these are far less easily observed and perhaps for the most part lethal, since MTOCs are also involved in cell division. The *Chlamydomonas* system may be a promising beginning to the discovery of more linkage groups and more orientational mutations.

DID SPIROCHETE SYMBIONTS GIVE RISE TO MOTILITY ORGANELLES?

IF THE MTOCs as amorphous, irregular, unbounded areas do actually have DNA or RNA (a possibility not ruled out, but still in contention), a symbiotic origin is by no means assured. Unlike the mitochondria and plastids, which retain so much of their bacterial heritage and which are conveniently contained in membranes, the MTOCs present a rather undefined starting point with which to develop a hypothesis of symbiogenesis. Fortunately for symbiotic theorists, there is something more (and something, by contrast, quite aesthetic) on which to develop further aspects of the theory.

Some fascinating symbiotic associations are known to exist between spirochetes and various eukaryotic hosts. Spirochetes are thin, squiggly, and highly motile bacteria that were first observed by A. van Leeuwenhoek from a sample of his own feces: "I have also seen a sort of animalcules that had the figure of our river eels. . . . Then had a very nimble motion, and bent their bodies serpent-wise, and shot through the stuff as quick as a pike does through water" (Dobell 1958). In associations with eukaryotes, spirochetes are found in large numbers attached to the host surface and with energetic and often concerted movement propelling the host forward.

The most famous of these symbioses is that of *Mixotricha paradoxa*, a mastigote which inhabits the intestine of *Mastotermes*, an Australian termite (Cleveland and Grimstone 1964). (See figure 8.5.) In this most dramatic of the known motility symbioses, approximately a half million small *Treponema*-like spirochetes, beating in synchrony, act as the motility system for the host (Cleveland 1956).[2] *Mixotricha* itself does have four motility organelles at its anterior end, and for this reason it is generally classified as a polymastigote. However, these four organelles probably do little for the motility of the cell and are hypothesized to act as rudders (although the ability of *Mixotricha* to change directions is rather limited). Each of the half million symbiotic spirochetes is attached at a specific holdfast site and alternates with an attached rod-shaped bacterium. The holdfast site appears under the electron microscope to have a specific morphology formed both by the attached tip of the spirochetes and the modified surface of the host. More casually attached, without a distinctive holdfast, are larger *Pillotina* spirochetes which probably do not contribute to the motility of the host. The smaller spirochetes beat in synchrony and thus confer a stately forward movement to *Mixotricha*, but the synchrony turns out not to be particularly unique in crowded populations of spirochetes. Blooms of free-living spirochetes attached to bits of debris may also be seen to beat in synchrony (Grosovsky and Margulis 1982); apparently it is easier to move synchronously if in intimate proximity to moving neighbors, and thus it is a convenient preadaptation for efficient motility symbioses.

[2] Film, however, remains one of the few methods by which this symbiosis has been elucidated, as *Mixotricha* is difficult to study. A similar and more accessible motility symbiosis of the termite organism called *Rubberneckia*, which is coated with flagellated bacterial symbionts, has been demonstrated by Tamm (1982).

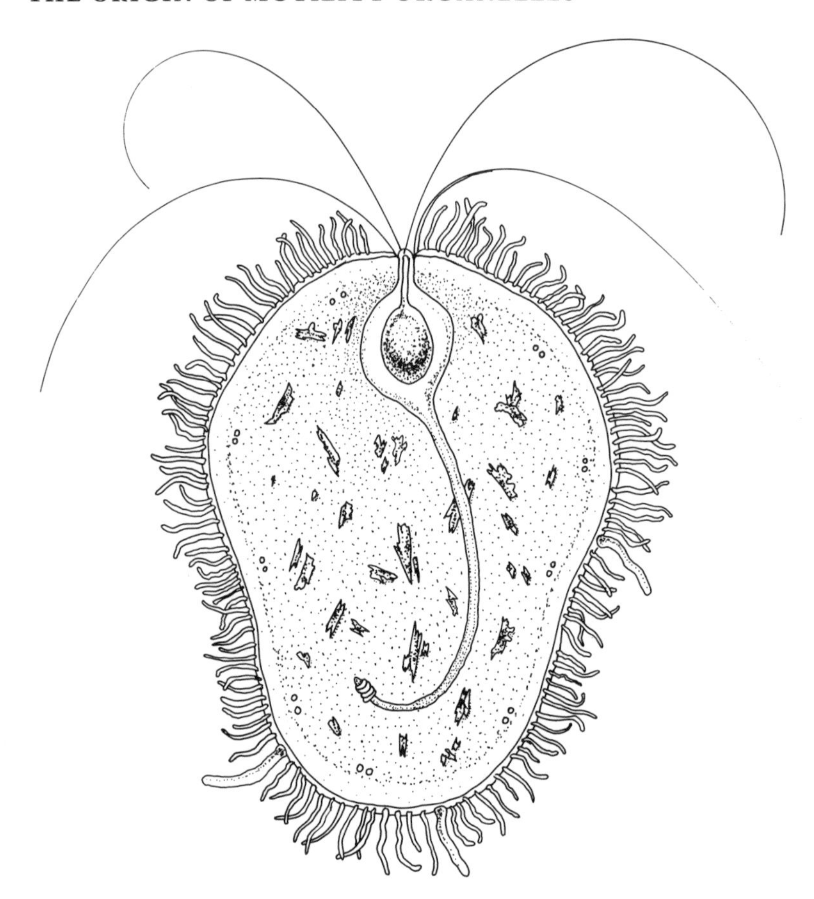

FIGURE 8.5. *Mixotricha paradoxa.* Drawing by C. Nichols.

If the origin of motility organelles of eukaryotes owes to a motility symbiosis with spirochetes, that hypothesis ought to be supported by predictions about the specific nature of spirochetes. For example, certain proteins, such a tubulin, the major component of motility organelles, could have been present in an ancient symbiotic spirochete, and therefore may also be found in some modern-day spirochetes. Indeed, the search for tubulin in spirochetes has had a promising start. Margulis et al. (1978) reported that *Pillotina* and *Hollandina* spirochetes from the termite *Kalotermes* could be stained with fluorescent antibody made against bovine, porcine, and sea urchin tubulin. Also "skinny, flagellated rod" bacteria were positive for the antibody. *Hollandina* spirochetes and skinny flagellated rods within the termite *Pterotermes* were also positive, while another spirochete *Treponema reiteri* was not. Miscellaneous small bacteria and some small spirochetes were negative. Although these results are not conclusive, the work was "state of the art" at the time. The researchers were encouraged to continue their research for a spirochete (or other bacterial) tubulin.

Another tantalizing piece of evidence is that transmission electron micrographs have revealed "microtubules" 20–24 nm in diameter both in cross section and in longitudinal section in *Pillotina* and *Hollandina* spirochetes from termites *Reticulitermes* sp., *Kalotermes* sp., and *Pterotermes.* These tubules are of unknown composition and structure and are not seen regularly in all sections. Some *Treponema*-like spirochetes have small tubules (Margulis et al. 1978); the spirochete *Leptonema* has even smaller tubules (Hovind-Hougen 1976).

Fracek and his colleagues (Bermudes et al. 1987) found that two proteins could be isolated from *Spirochaeta bajacaliforniensis,* using essentially the same technique commonly used to isolate α and β tubulin from mammalian tissue. Tubulin, while it is not unique in this regard, does have an unusual isolation method in which tissue is processed through cycles of warm and cold temperatures. During the cold cycles, tubulin depolymerizes and may be found in the supernatant of a centrifuged sample. During the warm cycles, tubulin is polymerized into tubules and will be found in the pellet. Most other proteins are "lost" during the cycling procedure, and nearly pure microtubules remain in the warmed pellets.

The next steps by which the two *Spirocheta* proteins were analyzed yielded ambiguous and, in one case, negative results. One of the two spirochete proteins was found to fluoresce with anti-α tubulin. This protein was sequenced and cloned (R. Obar, G. Tzertzinis and R. Laursen, unpublished results) and found to be a member of the hsp 60 heat shock family of proteins with a probable function of folding other proteins. The other spirochete protein (the one that did not fluoresce with anti-tubulin) has yet to be sequenced.

The *Hollandina* and *Pillotina* spirochetes, which showed both strong fluorescence with anti-tubulin and microtubule structures in their cytoplasm, have not been studied further because they cannot yet be cultured without the presence of other organisms. A next logical step is to use the new polymerase chain reaction (PCR) to search for tubulin-like proteins in partly pure preparations of *Hol-*

landina and *Pillotina*. The PCR method does not require that the organisms be in pure culture.

An investigation of another (apparently very common) eubacterial protein has yielded promising results. The filament temperature sensitive (ftsz) protein is coded by a gene that in *E. coli* includes a sequence that is a close match to a crucial and distinctive sequence of a tubulin gene (the GTPase binding site). FtsZ proteins are filaments that are assembled in microtubule or microfilament fashion and used to constrict the bacterial cell during division (RayChaudhuri and Park 1992; de Boer et al. 1992; Obar 1993). If ftsz protein is the ancestor of tubulin, then it is interesting that it seems to be found in eubacteria in general. This finding would, therefore, open the hypothesis of a symbiotic origin of motility organelles to bacterial symbionts beyond the spirochete group.

A PHYLOGENY FOR MICROTUBULES

THE PHYLOGENY of microtubule-based structures can be investigated on two levels—that of the proteins and that of the structures themselves. Three closely related forms of tubulin that are present in all eukaryotes are fundamental to microtubules. These forms are α- and β-tubulin, which polymerize in a 1:1 ratio to form the walls of microtubules, and γ-tubulin, which is apparently present only in MTOCs, where it is believed to anchor the microtubule projections. The specific γ-tubulin may in fact perform much of the "organizing" function of the MTOC. Analysis of the amino acid sequences of all three kinds of tubulin from fungi, animals, and plants shows the three to be equally distant from one another, leading to the conclusion that they all diverged from a common ancestor at about the same time. These tubulins are also highly conserved proteins, which gives the sequencing analysis more weight. The ubiquity of the three tubulins in eukaryotes places their common ancestor (the "missing link" for tubulin) no later than the earliest eukaryotic lineage. But where did this ancestor come from? Why did its early descendants diverge, and why was it lost?

A discussion of the evolution of microtubule functions may help to address this question. Four distinct microtubule-based organelles are widespread in the eukaryotic kingdoms, but are absent from extant bacteria. These organelles are: the MTOC, the motility organelle, the spindle microtubules of mitotic and meiotic division, and the cytoplasmic microtubules (which are both tracks for intracellular transport and an essential component of the cytoskeleton). The pattern of occurrence of these microtubule-based structures indicates that they arose suddenly and radiated within a very short period in the earliest eukaryotic lineage.

The functions of three of the four microtubule-based systems—cell motility, cell division, and cytoskeletal transport—involve motility based on motor proteins (usually members of the dynein protein superfamily). The dynein proteins are ATP-driven mechanochemical motors that have molecular weights greater than 200,000 daltons. Spindle function and intracellular transport also rely on members of the kinesin protein superfamily, another large group of ATP-driven motors.

These observations lead us to propose a preliminary phylogeny of these proteins (figure 8.6). An early branching pattern for functions of the MTOCs would make them available for the elaboration of the complex motility and spindle structures that depend on them. This branching pattern may have been facilitated by a diversity of motor proteins (and other associated proteins) that evolved in eukaryotes. A circumstantial point supporting this scheme is that bacteria rarely synthesize proteins larger than 100,000 daltons. (Bacterial polymerases may be as large as 150,000 daltons, but polymerases are properly viewed in terms of their smaller units.) Biosynthesis of a typical motor protein would take several minutes, a long time in the life of a bacterium, and these proteins probably could not have evolved in other than a eukaryotic setting. An intriguing alternative is that motor proteins might have evolved initially in Archaebacteria, the proposed source of the nucleocytoplasm of eukaryotes. Archaebacteria do grow slowly and thus may have the time to make large proteins.

The absence of all four MT-based structures from bacteria, and their ubiquity in eukaryotes, implies sudden acquisition of one or more genes by an early eukaryote (i.e. already containing mitochondria) followed by a rapid evolution of specialized microtubule structures. This pattern is a hallmark of a symbiotic association, which is discussed in more detail next.

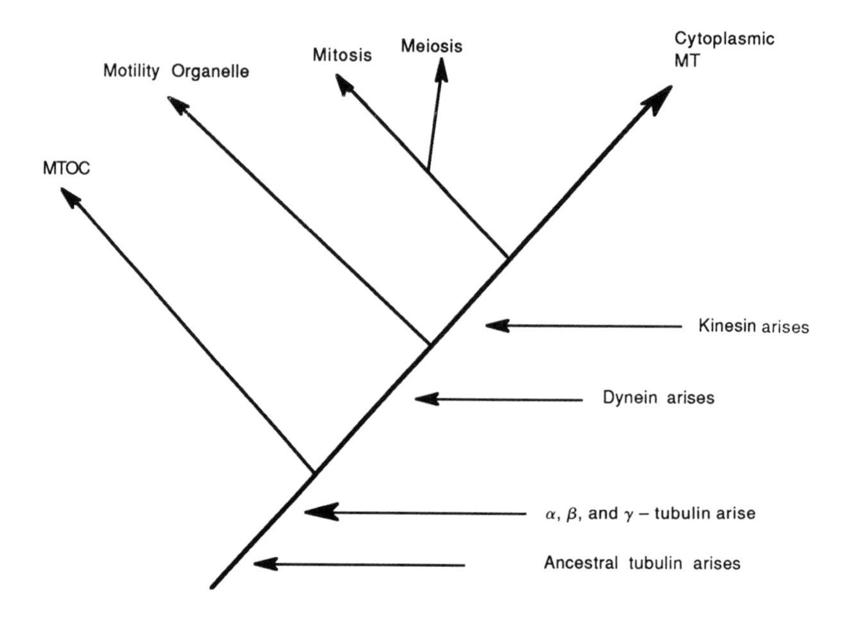

FIGURE 8.6. A hypothetical phylogeny for microtubule associated proteins.

A SCENARIO FOR THE ACQUISITION OF MOTILITY SYMBIONTS

THE FOLLOWING scenario outlines the steps that might have occurred if eukaryotes or pre-eukaryotes acquired tubulin-bearing spirochetes as motility symbionts:

1. Microtubules made of a tubulin-like protein evolved in free-living spirochetes. The function of these microtubules, which may be observed in some modern-day spirochetes, is unknown (figure 8.7). It is fairly certain that the function is not directly involved with motility, as spirochetes already have an adequate motility system involving bacterial flagella. Is there an indirect motility function? Some small microtubules of treponemes are attached to the basal structures that cause the bacterial flagella to move. Might they form a part of the cytoskeleton of some spirochetes, similar to one of the microtubule functions in eukaryotes?

2. Spirochetes (with microtubules and tubulin) next formed sym-

biotic associations with pre-eukaryotic or early eukaryotic host cells. As can be observed in many extant motility symbioses, the relationship featured a distinctive holdfast site formed by both the modified end of the spirochete and the modified membrane of the host. Extant examples may be seen in figure 8.8.

Just as mitochondria and plastids share coding with the nucleus for certain crucial gene products, a genetic transfer to the nucleus probably happened with the spirochete symbionts too. Shared coding is a mechanism by which partners in an association maintain mutual control, each preventing the other from outgrowing the relationship. A motility symbiosis differs from the more internal symbiosis of mitochondria and plastids, however, in at least two crucial ways. First, the entire spirochete is not in close physical contact with the host, only the holdfast end is. Second, the major function of the relationship is not the assimilation of energy but actually the consumption of energy during movement. It is true that movement may propel the entire complex toward a food source or sunlight but the direct consequence of a motility symbiosis is probably a drain of energy, albeit a worthwhile drain. Therefore, it is unlikely that the same sequence of events that apparently occurred in both mitochondria and plastids would also be responsible for the acquisition by eukaryotes of motility organelles.

3. Once the symbiont was attached, the number of genes carried by the spirochete symbiont would have decreased markedly—a loss that echoes the symbiogenesis of mitochondria and plastids. Lost would be any genes that would confer a selective advantage only to a free-living bacterium. Genes necessary for fermentation (the presumed form of metabolism for both host and symbiont) would have vanished from their redundant location in the spirochete. In fact, there is something in the way that extant motility symbioses develop that suggests that fermentation might have been lost quickly.

Symbioses that form in termite guts provide the clue. Many free-living spirochetes appear to be grazing on products exuded from mastigotes in the termite intestine. Some spirochetes form casual associations, not conferring motility, but presumably taking advantage of some rich exudate. A next step, firm attachment and motility, would probably not bring with it any advantage in the form of more fermentation to the host, as the host was apparently already feeding the symbionts. Some genes, nevertheless, by chance would likely be transferred to the nucleus by the same mechanism believed

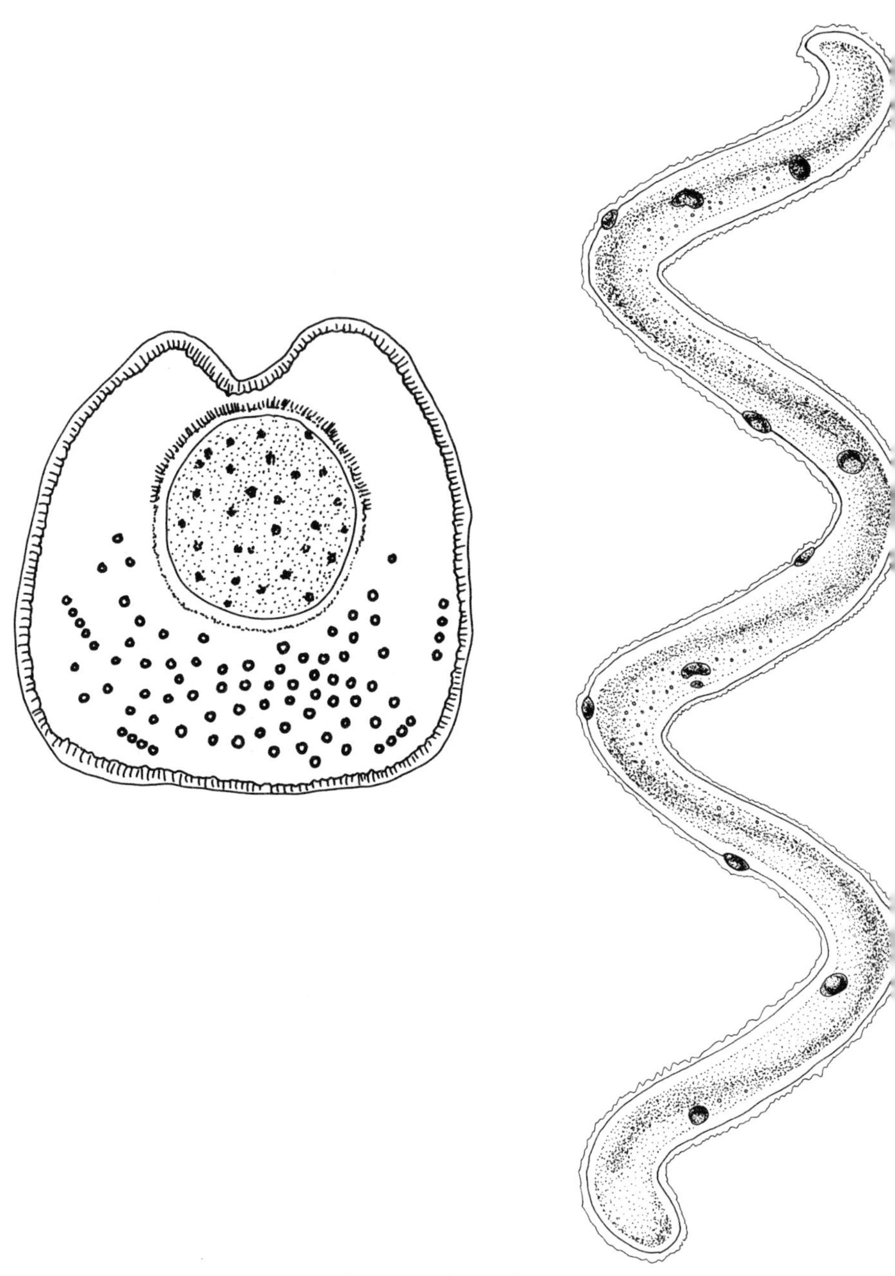

FIGURE 8.7. a. Cross section of a spirochete showing bacterial flagella (larger circles) and tubules (smaller circles in center). Drawing by C. Nichols. b. Spirochete. Drawing by C. Nichols.

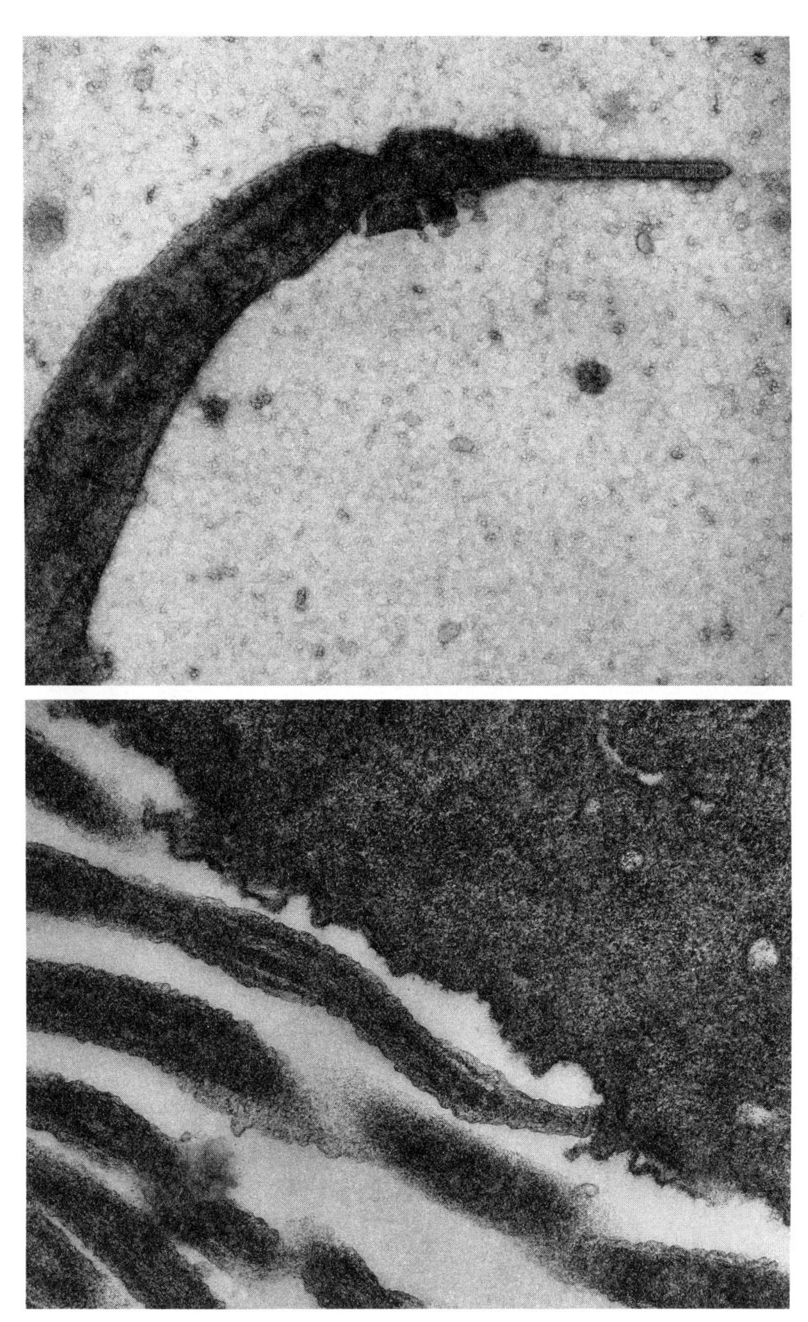

FIGURE 8.8. a. Spirochete holdfasts from protist of *Inscisotermes* (a) and *Reticulitermes* (b). Photos by David Chase.

to have streamlined the mitochondria and plastids. All of this gene transfer and loss would leave the spirochete with a rather small genome.

Theorists of the symbiotic origin for mitochondria and plastids have been fortunate that there are remnants of genetic autonomy left in these organelles, otherwise the primary evidence for symbiosis would not exist. The problem for motility organelles—specifically for MTOCs—and for other suspected products of symbiogenesis (hydrogenosomes and microbodies or peroxisomes) is that only a few functions of the hypothesized genome remain, or even none at all. If gene transfer has occurred, then certainly one of the primary genes acquired and used extensively by the nuclear genome was the gene for tubulin, which produces an extraordinarily versatile and useful product.

Tubulin has a number of unique properties that might have preadapted it for the diverse functions it now serves in eukaryotes. Tubulin can self-assemble into microtubules, under many ordinary cellular conditions and in the test tube. Microtubules readily bind to an enormous array of other molecules, ranging from the very simple polylysine (just a repeated polypeptide chain of the amino acid lysine) to complicated motor molecules. Many molecules in the early eukaryotic cell could have attached to microtubules and a significant few (microtubule associated proteins or MAPs, such as the molecular motors) might have been preadapted to perform some particular function with microtubules. Some of these functions must have dramatically increased the efficiency of the host cell. For example, microtubules as cytoskeletal structures would have helped the cell maintain its form. Microtubules with MAPs would have facilitated the transport of vesicles and other structures inside the cell. Some of the structures so moved would have been chromosomes, and this would have enabled a form of cell division and chromosome separation (mitosis) so efficient that it is now almost universal in eukaryotes. Although the natural sequence of events in mitosis varies markedly among eukaryotic taxa, the transformation of microtubules into spindles is characteristic of all except the karyoblasteans.

4. At some point into the symbiogenesis, therefore, only a tiny spirochete genome remains after numerous redundant and unnecessary genes have been lost and others transferred by chance to the nucleus. The genome retains a function and position commensurate with its only remaining role: to maintain a relic of autonomy in the

symbiont. We would expect this genome to be present in the MTOC, which in a motility symbiosis would be the area of mutual control by the partners, the holdfast site. We would expect that whatever genes were left, some might have something to do with maintaining and controlling the most important part of the structure, the tubule. This, in the case of *Chlamydomonas,* is what the umi linkage group is in fact doing. The umi genes all encode products used in the MTOC that enable microtubule structures to function normally.

5. The remnant of the spirochete genome, now located in the MTOC area, may have been acquired without the convenience of a distinctive enclosure, like the membranes of mitochondria and plastids. The symbiosis, presumably, never was a truly internal one. Thus the only vestige of the original symbiont is a small amorphous area associated with microtubule structures that seems to contain DNA but which cannot be reliably isolated as a complete entity. It is the apparent semiautonomous self-replication of the MTOC, the hallmark of an independent genetics, that led originally to all of the subsequent research in this area.

SELECTION PRESSURES FOR ACQUIRING MOTILITY ORGANELLES

THE BEST-KNOWN examples of ongoing motility symbioses all take place in anaerobic environments, specifically the digestive systems of wood-eating roaches and termites. These symbioses consist of spirochetes or various other flagellated bacteria attached, sometimes quite specifically, to host protoctists. In some cases, the concerted movement of hundreds to thousands of undulating spirochetes propel the host forward. Even other transiently observed motility symbioses, such as one observed by Dyer (unpublished) that entailed an unknown ciliate with spirochetes attached at its posterior end, are from deep anaerobic sediments.

Some bacterial consortia, too, may be in part motility symbioses. For example, *Chloropseudomonas* was at first described as a single green bacterium that was rod-shaped and motile, but it turned out to be a symbiosis between *Chlorobium* (a nonmotile green bacterium) and a motile sulfate- or sulfur-reducing bacterium. Other green bacteria are regularly found arranged around a motile colorless bacterium (Stanier, Pfennig, and Truper 1981). These apparent motility symbioses, too, are in anaerobic environments. Do anaerobic environments somehow select for motility symbioses or is this coincidence?

One possibility is that the tendency for spirochetes and other bacteria to be found "grazing" around anaerobic protoctists might be a function of the relatively incomplete fermentative metabolism typical of anaerobes. The waste products of such metabolism are thus attractive foods for the surrounding bacteria. A bacterium feasting on the protoctist wastes would have an advantage in becoming firmly attached to the protoctist. Motility might have come as a fortuitous by-product of this interaction. Such a scenario is supported by the observation that many other bacteria that apparently are not motility symbionts become attached especially to anaerobic protoctists.

A second possibility is that there really is something about the anaerobic environment that selects for motility symbioses. This would explain why some grazing spirochetes became not only firmly attached but also precisely arranged in orderly arrays on the surface of a host, and why the attachment sites are often so elaborate. Movement requires ATP, and this energy-rich molecule is somewhat difficult to make with anaerobic metabolism. For example, only two ATPs are formed in glycolysis of a single glucose molecule into pyruvate while thirty-eight ATPs are formed when a single glucose molecule is respired aerobically (in mitochondria) in animal cells. At least one ATP is used up each time one dynein molecule interacts productively with a microtubule in a motility organelle. There are thousands of dyneins in each motility organelle and each motile stroke requires each dynein to use up ATP. An anaerobic environment, therefore, may select for concerted movement with more efficient use of ATPs.

The schooling behavior of some fish is a useful analogy. The fish are known to arrange themselves at diagonals to each other so that the currents generated by the tail fins of neighboring fish can be used optimally for propelling each individual forward (Weihs 1973; Walker 1977:260–61). Long distance runners and swimmers "draft" off each other in the same way. A similar mechanism may be occurring in symbiotic spirochetes, which are often arranged at specific diagonals to each other on the surface of their host. Spirochetes and most other small aquatic organisms have the additional problem of lack of traction in an aqueous environment. Firm attachment to a host, combined with the tendency of spirochetes in close proximity to beat in synchrony (apparently the most energy efficient way to move under crowded conditions), may help to solve a major problem of microscopic motility. That is, for organisms of this size water is as viscous as glycerin would be for a human swimmer. Any edge in motility might well confer an important advantage in a microorganism, especially one struggling to make ATPs.

Tamm (1982) studied a motility symbiosis of a devescovinid "Rubberneckia" (from a termite) and its flagellated bacteria. The independently moving bacteria apparently attain speeds of up to 50–60 micrometers (μ) per second. The bacteria attached to the host propel it at speeds of 100–150μ per second. One possible explanation, according to Tamm, is that an individual bacterium experiences a drag (friction) of about 1/40th that of its larger host. However, there are about 2,500 bacteria attached to the host, and thus the drag per bacterium is actually 2,500/40, or about 60 times less than that for an individual. This enables the bacterial flagella to rotate faster and to propel the host more efficiently.

We propose therefore that the selection pressure for the acquisition of motile spirochetes, which must have happened well before mitochondria and plastids were acquired, was the limited efficiency of ATP production in an anaerobic environment. Motility symbioses would have enhanced the efficiency of movement toward food, away from oxygen, toward sunlight, etc.

FUTURE RESEARCH

ALTHOUGH TUBULIN and tubulin structures are one of the defining characteristics of all extant eukaryotes, motility organelles themselves are not. We hypothesize that motility organelles were secondarily lost in nine of the thirty-six phyla of protoctists, most of which are amoebae or obligate parasites. Red algae and conjugating green algae also have lost motility organelles. Another possibility is that in some of these groups, such as the amoeboid ones, spirochetes were acquired directly as endosymbionts; the interior of a cell is a typical environment for many extant spirochete pathogens. In this way, the external motility symbiosis might have been bypassed. If tubulin was acquired via both paths, then the symbiogenesis of this aspect of the eukaryotic cell would have been polyphyletic.

Ribosomal RNA data may provide clues by showing presence or lack of ancestors with motility organelles for some of these phyla. Amoeboid organisms, for example, seem to be quite polyphyletic, and the story underlying a putative loss of motility organelles may be different for each one.

The origin of tubulin structures in eukaryotes remains one of the more controversial and elusive problems at the frontier of cell evolution. We ourselves favor symbiotic origin hypotheses for tubulin structures, but we readily acknowledge that the symbiogenesis of tubulin structures—in contrast to the symbiogenesis of mitochondria and plastids—is in no way yet substantiated. A more thorough search for tubulin and tubulin-like proteins, as well as for proteins associated with tubulins, must be undertaken in both archaebacteria and eubacteria. Those intriguing observations of cortical genetics, the uni group, and motility symbioses must be more clearly understood. Interesting years lie ahead for any researcher who cares to tackle a part of this intricate problem.

IX

Outstanding Questions and Directions for Future Research

WHAT INTERESTING times these are for researchers studying the origins of eukaryotes! New techniques in molecular evolution, such as sophisticated computer software and the polymerase chain reaction, now enable the many sequencing studies needed for this work to be done efficiently and even somewhat inexpensively. Just a short time ago it would have taken years to generate and then disseminate some of the sequence data that now takes just days or weeks.

The juxtaposition of two seemingly disparate fields, paleontology and molecular evolution, is one of the most important advances in interpretation of the data and observations collected in both these fields. The details of evolution revealed by such a confluence of studies, such as Carl Woese's work on prokaryotic taxonomy, is almost overwhelming in their complexity. On the other hand, it is with relief that we see some branches of the family tree displaying a simple logic—such as the lack of mitochondria in early eukaryote branches, which confirms that these branches diverged before the remaining trunk of the early eukaryotes acquired such organelles.

The many, many outstanding questions about eukaryotic evolution will provide decades of exciting research in this field. There are details to be sorted out, obscure groups of protoctist and prokaryote molecules to be sequenced, and yet more fossils to be discovered.

While assembling this book we contacted some of the researchers interested in eukaryotic evolution,[1] and asked them what they thought were the important questions for research focus. Their replies, along with many of our own ideas, are organized here according to the chapter titles in this book. We found this exercise not only useful, but inspiring; it makes us want to get right back into the laboratory to tackle one or two of the problems.

Fossils and Molecules

The fossil record for eukaryotes is sketchy, especially prior to about two billion years ago. The search, therefore, must strive to fill in as many gaps as possible. Further, some microfossils and megafossils already discovered and stored in collections around the world may need reinterpretation in light of rapidly accumulating data from the molecular evolutionists.

Robert Horodyski suggests that new techniques may need to be developed to recognize megafossils (which may be more common than once suspected), such as *Grypania*. Was the mid-Proterozoic a time for experimentation with giantism? Perhaps we should we be looking for uncommonly large cells such as cellular slime mold plasmodia, networks of fungi such as *Armillaria,* phaeophytes like kelp, or even massive assemblages of myxobacteria. And, Horodyski

[1] Peter Beamish, Ceta Research, Inc., Trinity, Newfoundland, Canada; Jacques Berger, University of Toronto, Ontario, Canada; W. Ford Doolittle, Dalhousie University, Halifax, Nova Scotia, Canada; Michael Enzien, Westinghouse Savannah River Co., Aiken, S.C.; Nicholas Gillham, Duke University, Durham, N.C.; Stephen Giovannoni, Oregon State University, Corvallis; Ricardo Guerrero, Universidad de Barcelona, Barcelona, Spain; Patrick Gunkel, Ideonomist, Austin, Texas; J. Woodland Hastings, Harvard University, Cambridge, Mass.; Robert Horodyski, Tulane University, New Orleans, La.; Kwang Jeon, University of Tennessee, Knoxville; Wolfgang Krumbein, University of Oldenburg, Oldenburg, Germany; Ralph Lewin, Scripps Institute of Oceanography, La Jolla, Calif.; Alejandro Lopez-Cortez, Centro de Investigaciones, Biologicas de Baja California, Mexico; Mark McMenamin, Mt. Holyoke College, South Hadley, Mass.; Jeffrey Palmer, Indiana University, Bloomington; K. A. Pirozynski, Canadian Museum of Nature, Ottawa, Canada; Dennis Searcy, University of Massachusetts, Amherst.

adds, a greater variety of cherts should be examined, not just those from hypersaline settings.

Wolfgang Krumbein wonders why, after such a successful reign by prokaryotes, did eukaryotes evolve? And do eukaryotes participate differently from prokaryotes in biogeochemical cycles?

Is there a connection, Horodyski adds, between the cooling of the Earth in the mid-Proterozoic and the evolution of eukaryotes? What other signatures might there be for the evolution of eukaryotes? More should be done to establish reliable chemical or biological markers. Are there any particular signatures for symbiosis in the fossil record, such as changes in size, morphology, behavior, and physiology?

Finally with respect to the fossil record, after the establishment of eukaryotes might the trend in cell size have gone from large to small, reflecting increased efficiency and coordination of the partners?

With respect to molecular evolution, more work on sequences and with more organisms is needed, as the number and branching patterns of the main branches of organisms are still in dispute. Also, algorithms for data analysis could be refined further to take into consideration differences in rates of evolution.

Dennis Searcy wonders what might account for the differences between Archaebacteria and Eukaryotes, which in spite of some striking similarities have evolved for the most part along very different pathways. Stephen Giovannoni suggests that there might be primitive eukaryotes as yet undiscovered—a safe bet considering the tiny fraction of eukaryotes with rRNA sequences analyzed.

The Evolution of Metabolism

More and more bacterial sequences would help to confirm and elaborate the already intricate family tree developed by Carl Woese. This more fully developed family tree must then be thoroughly analyzed for what it can tell us about the evolution of metabolism. Are there remnants of genes for lost metabolisms in various bacterial groups (such as remnants of ancestral photosynthesis)? Can we sort out the differences between retention of semes and horizontal transfer? And how did some of the truly fundamental cycles such as the Calvin and Krebs cycles evolve?

How much can we rely upon the sometimes enigmatic fossil record to provide (or confirm) dates for the emergence of various types of metabolism. Are there any particular signatures for types of metabolisms not yet discovered or recognized?

Horizontal Gene Transfer

Yet to be determined is just how extensive and evolutionarily important horizontal transfer might be. Are the numerous documented examples just the tip of the iceberg? Are we all intricately pieced-together chimeras?

The Eukaryotic Host Cell

More protoctist sequences will always be welcome, say Ralph Lewin, W. Ford Doolittle, and many others. For example, karyoblasteans, retortomonads, and pyrsonymphids (all kinds of protoctists) do not yet have a place on the family tree. And might there be eukaryotic branches even closer to the Archaebacterial line? Those eukaryotes that are currently considered to be most primitive—the microsporans, diplomonads, and parabasalids—are parasites (or symbionts); which of their traits are actually primitive and which are advanced?

Can an extant prokaryotic group (presumably archaebacterial) be found that does anything like exocytosis or endocytosis or amoeboid movement?

And what, ask many of our respondents, are the origins of the many complex parts of the eukaryotic cell: the internal membranes, the nucleus, the cytoskeleton, the various organelles and membrane-bound structures. Mitochondria and plastids are about the only eukaryotic cell components with well-understood genealogies.

The story of the evolution of introns needs more data and analysis before it can be considered minimally coherent. While a few types of introns seem to provide a connection between Archaebacteria and Eukaryotes, most types just seem to confuse the story. Also, the current classification system for introns is becoming obsolete, as more introns are discovered that do not quite fit. A new classification system should attempt to take evolution into consideration.

What other types of symbioses might be useful in understanding eukaryotic cells? Are there sensory symbioses in which signals are transmitted from one partner to another? Are there symbioses based upon defensive behaviors, such as the kappa particles (bacteria) of Paramecium?

J. Woodland Hastings wonders how circadian rhythms (biological clocks) evolved and what their connection might be with symbiosis. Circadian rhythms have been found (so far) only in eukaryotes.

Is there a connection between the acquisition of symbionts (such as mitochondria) and the evolution of multicellularity, differentiation, aging, and death? And what principles of population biology (on a cellular level) and symbiosis might help in understanding the pathologies of the eukaryotic cell?

The Evolution of Eukaryotic Sex

The discovery and study of more mutants for specific aspects of sexuality will help elucidate the origins of meiotic division and synapse formation. L. R. Cleveland worked out the details of sexuality for zoomastigina, thus revealing several possible links in the evolution of meiosis. Are there other protoctists for which such an analysis might be revealing?

What are the selection pressures for the evolution of sex, the maintenance of sex, and the secondary loss of sex? Is sexuality polyphyletic in eukaryotes?

The Acquisition of Mitochondria

The Mitochondria have perhaps the best understood evolutionary history of all the structures in the eukaryotic cell. Nevertheless, numerous questions remain. Are mitochondria polyphyletic? Are there degenerate mitochondria such as, perhaps, the hydrogenosomes? Have secondary losses of mitochondria occurred? Where would one look for evidence?

Mitochondria have retained parts of their original genomes, but why have they retained any at all? What determines whether a gene remains or is lost or transferred? The mitochondria are nearly unique in having a slightly different genetic code than the universal one. Why did this occur in mitochondria and (apparently) in no other organelles? To what extent is this phenomenon of code change predictable?

Dennis Searcy wonders whether stable mitochondria-like symbionts could be assembled in the lab, and Kwang Jeon asks whether conditions might still be favorable for an ongoing establishment of new symbiotic organelles.

THE ACQUISITION OF PLASTIDS

IT APPEARS that plastids are polyphyletic symbionts; the clues lie not only in sequence data but in the presence of multiple plastid membranes and specific pigment signatures. However, many respondents would like to see more evidence and analysis.

Nicholas Gillham asks why (if mitochondria and plastids are on separate lineages and plastids are polyphyletic) do mitochondria and plastids carry such similar proteins in their electron transport and ATPase complexes?

Some plastids and their genomes are known to be quite minimal, such as those of parasitic plants. Can plastids or their genomes be entirely lost in some circumstances? Jeffrey Palmer wonders whether there are other heterotrophs (such as apicomplexans) that harbor minimal plastid genomes. Might there be plastids so disintegrated as to be unrecognizable, such as in the watermolds oomycetes? asks Mark McMenamin.

THE CONTROVERSY ABOUT THE ORIGIN OF MOTILITY ORGANELLES

NONE OF our correspondents said that they were convinced that a symbiotic origin for motility organelles has been demonstrated (although some find the hypothesis intriguing). More data and observations are needed. Most importantly, there should be a relentless search of the prokaryotes for tubulin-like proteins. If such proteins should turn up in Archaebacteria, that might suggest an endogenous origin of microtubule structures in the eukaryotic cytoplasm. If found in Eubacteria, it might suggest an acquisition via symbiosis. And where do those functionally important MTOCs come from?

Whether there is DNA in the MTOCs is yet to be sorted out. The uni-linkage groups and ciliate cortex genetics must be studied further. And one could return to the difficult problem of greatly reduced remnants of symbioses, seen in some plastids and perhaps in hydrogenosomes and microbodies. Could a structure, under any circumstances, be deemed of symbiotic origin if there were no remnant of a genome left?

What more can extant motility symbioses tell us? Is the schooling fish analogy pertinent for these symbioses? Is there a connection between the evolution of sensory systems and the evolution of tubulin and motility systems?

WE (WITH the help of our respondents) have just asked enough questions to keep hundreds of researchers and theorists busy (and presumably happy) for hundreds of thousands of hours. Indeed, even as we finish this manuscript, papers are coming out that will confirm, complicate, or refute some of the statements that we have made in this book. But this research foment is to be hoped for; it is indication of the general well-being and productivity of research on the evolution of eukaryotic cells. For those who love a challenging quarry, the grin on this one should appear irresistibly enigmatic.

APPENDIX A

What To Do About Taxonomic Terminology

SCIENTIFIC TERMINOLOGY enables workers in a field to communicate quickly and accurately with each other. Students of science often find learning the necessary terminology as arduous and yet as ultimately rewarding as studying a foreign language. One hopes and strives for terminology that not only is accurate and useful but fits some overall organizational scheme such as (and especially) evolution. Indeed, good terminology should help one to *see* the overall picture. This is why, in the long run, rRNA and other sequence data will be valued for their help in making vast improvements in taxonomy, which will eventually be reflected in the taxonomic terminology. We agree with Woese (1987) who concludes that the sooner the old taxonomy is adjusted to reflect the new data, the better. It will be the next generation of researchers who will really use, appreciate, and understand the new sequence-based system.

However, we do not agree with Woese (and others) who judge that the established Five Kingdom system needs to be entirely scrapped. Ousting the Two Kingdom system (Animals and Plants) from the textbooks and classrooms was a long, difficult battle. The Five Kingdom system now in textbooks at all levels is taught in most biology classes and is even reflected in the organization of catalogues for biological supply houses. This classification system has helped to bring out of the shadows some of the obscure eukaryotes, such as many of the protoctists and fungi. It has placed the four eukaryotic Kingdoms into an evolutionary perspective (and one that is still accurate based on the sequence data as well).

For the purposes of researchers in evolutionary biology, the three oldest branches of organisms—Archaebacteria, Eubacteria, and Eucarya (giving rise to the Eukaryotes)—should be considered superkingdoms, with the four eukaryotic kingdoms maintaining their kingdom status, but now within the superkingdom Eucarya. However, the average elementary, middle, and high school student and even beginning-level college students, not to mention their teachers, have been greatly aided in the understanding of cell evolution by the Five Kingdom system. Retaining the Five Kingdom system for most practical purposes should not be regarded as an attempt to "dumb down" the terminology but to help students and teachers who otherwise have little appreciation (or even knowledge of the existence) of bacteria, protoctists, and fungi—let alone their significance in evolution. Dividing the bacteria, obscure as they are in the experience of the average person, into two kingdoms and leaving all of the eukaryotes lumped together as one (with the additional possibility of assigning higher classifications to obscure, dangling groups like the microsporans) is of no real help—although it might satisfy desires by some researchers in molecular biology who would like absolute concordance of classification schemes with known (and likely to keep changing) hierarchical relations.

The molecular biologists should perhaps step back from their data a little (compelling as they are) to ask whether it might not make sense to give the serial acquisition of symbionts at least as much taxonomic weight as the accumulation of a few dozen mutations in a conserved molecule. The eukaryotic kingdoms are not just tiny branches at the tips of the family tree, although they are sometimes pictured that way by an artifact of sample size (e.g., using mice and fruit flies to represent all animals). Sequence data are important but not so important that such data should be allowed to wag the entire dog. Other criteria, such as endosymbiosis, must be considered as well. Furthermore, the tendency for symbiotic organisms to exchange genes may homogenize sequence data, perhaps giving a false appearance of a slower and less divergent evolution in the eukaryotes—which, after all, share many of the same internal symbionts. The apparent extreme divergence between, for example, Archea and Eubacteria, may be a consequence less of a vast stretch of time since their divergence and more a consequence of the extremely different habitats in which their constituent species have lived their lives, thereby lessening opportunities for exchange.

Overall, we do not consider the Five Kingdom system to be as fraught with problems as Woese (1987) and others contend. But at the same time, we welcome the addition of three superkingdoms (not to mention the almost total rearrangement of bacterial taxonomy).

The real problem, as we see it, is that many remnants of the old two-kingdom system still exist in the literature as terminology that should be vanquished. Unfortunately, some of the little branches of protistology, mycology, botany, and zoology have lexicons that make them, literally, kingdoms unto themselves. Terminology is used like passwords for entrance into the castle. Margulis et al. (1989) made a valiant attempt to clean up much of this terminology in their *Handbook of Protoctists.* However, they were met with considerable resistance, and so even in that book much of the redundant and obscuring terms remain.

APPENDIX B

Is It a Symbiont?

CRITERIA FOR identifying a particular organelle or structure in a eukaryote as being of symbiotic prokaryotic origin should include:

1. enzymes and protein complexes within the organelle are more similar to prokaryotic than to eukaryotic cytoplasmic analogues.
2. a free-living prokaryote can be found that has strong genetic, biochemical, and morphological resemblances to the organelle.
3. the organelle itself retains a genome, with characteristics more similar to those of a prokaryotic than a eukaryotic genome.
4. ribosomal RNA, transfer RNA, and messenger RNA within the organelle are more similar to those of prokaryotic than of eukaryotic cytoplasm.
5. the organelle has an ability to replicate and a genetics that is separate from the nuclear genetics.
6. the organelle is found in eukaryotes as an "all or nothing" phe-

> nomenon in which one either finds the organelle as a whole (or in some modified form) or does not find it at all (if it has not been acquired or was secondarily lost); one does not expect to find intermediate stages of the organelle if it was acquired all at once as a symbiont.

This should be taken as a "wish list"; although mitochondria and plastids seem to be exemplary cases of symbiotic origin, many other structures will most likely fall short of these criteria, having secondarily lost most of their capabilities.

APPENDIX C

Steps for Horizontal Gene Transfer in Symbioses

I. The prerequisites for successful gene transfer

of *protein coding genes* are mechanisms for transcription, translation, post-translational processing, and transport of gene product to its functional site;
of *introns* are mechanisms for transcription and excision from the primary transcript (in the case of group I introns that encode a maturase, translation is also required); and
of *rRNA and tRNA genes* are mechanisms for transcription, post-transcriptional processing, and transport to functional site.

II. The sequence of possible steps in Productive Gene Transfer between genomes *within a single organism* are

1. Positive selection for the function or product of the gene should preexist.

2a. Duplication or polyploidy of the gene to yield a DNA copy for transfer, or
2b. Transcription of the gene to yield an RNA copy for transfer (either a DNA or RNA copy will do as long as an "archival" DNA copy is left intact).
3. Transfer of a copy of the gene.
4. Activation of the transferred copy through acquisition of a leader (signal) sequence for organellar protein, or acquisition of genetic elements needed to regulate transcription in the recipient cell (e.g., introns, promotors, enhancers).
5. Selection between retained and transferred copies of gene.
6. Loss of retained copy of gene.

REFERENCES

Alberts, B., D. Bray, J. Lewis, M. Raff, K. Roberts, and J. D. Watson. 1983. *Molecular Biology of the Cell.* New York: Garland.

Awramik, S. M. and J. P. Vanyo. 1986. Heliotropism in modern stromatolites. *Science* 231: 1279–81.

Axelrod, R. 1984. *The Evolution of Cooperation.* New York: Basic Books.

Bakker, R. T. 1986. *The Dinosaur Heresies: New Theories Unlocking the Mystery of the Dinosaurs and Their Extinction.* New York: Morrow.

Barghoorn, E. and J. W. Schopf. 1965. Microorganisms from the late precambrian of Central Australia. *Science* 150: 337–39.

Barghoorn, E. and S. Tyler. 1965. Microorganisms from the Gunflint chert. *Science* 147: 563–77.

Barrell, B. G., A. T. Bankier, and J. Drouin. 1979. A different genetic code in human mitochondria. *Nature* 282: 189–94.

Becker, M. and G. Schaefer. 1991. Purification and spectral characteristics of the b-type cytochrome from the plasma membrane of the archaebacterium *Sulfolobus acidocalderius. FEBS* 291: 331–35.

REFERENCES

Belfort, M. 1991. Self-splicing introns in prokaryotes: Migrant fossils? *Cell* 64: 9–11.

Bermudes, D. and R. C. Back. 1991. Symbiosis inferred from the fossil record. In L. Margulis and R. Fester, eds., *Symbiosis as a Source of Evolutionary Innovation*, pp. 72—94. Cambridge, Mass.: MIT Press.

Bermudes, D., S. P. Fracek, Jr., R. A. Laursen, L. Margulis, R. Obar, and G. Tzertzinis. 1987. Tubulin-like protein from *Spirochaeta bajacaliforniensis*. *Annals of the New York Academy of Sciences* 503: 515–27.

Birky, C. W. 1978. Transmission genetics of mitochondria and chloroplasts. *Annual Review of Genetics* 12: 471–512.

Boer, P. H. and M. W. Gray. 1988. Transfer RNA genes and the genetic code in *Chlamydomonas reinhardtii* mitochondria. *Curr. Genet.* 14: 583–90.

Brugerolle, G. and J. Mignot. 1989. Retortomonadida. In L. Margulis, J. Corliss, M. Melkonian, and D. Chapman, eds., *The Handbook of Protoctista*, pp. 259–65. Boston: Jones and Bartlett.

Buchanan, R. E. and N. E. Gibbons, eds. 1974. *Bergey's Manual of Determinative Bacteriology*, 8th ed. Baltimore: Williams and Wilkins.

Butterfield, N. J., A. H. Knoll, and K. Swett. 1990. A bangiophyte red alga from the Proterozoic of Arctic Canada. *Science* 150: 337–39.

Canning, E. 1988. Nuclear division and chromosome cycle in microsporidia. *Biosystems* 21: 333–40.

Canning, E. 1989. Microspora. In L. Margulis, J. Corliss, M. Melkonian, and D. Chapman, eds., *The Handbook of Protoctista*, pp. 53–74. Boston: Jones and Bartlett.

Carpenter, A. T. C. 1979. Recombinational nodules and synaptonemal complex in recombinant-defective females of *Drosophila melanogaster*. *Chromosoma* 75: 259–92.

Castenholz, R. W. 1981. Isolation and cultivation of thermophilic cyanobacteria. In M. Starr, H. Stolp, H. Trüper, A. Balows, and H. Schlegel, eds., *The Prokaryotes*, pp. 236–46. Berlin: Springer-Verlag.

Castenholz, R. W. and B. K. Pierson. 1981. Isolation of members of the family Chloroflexaceae. In M. Starr, H. Stolp, H. Trüper, A. Balows, and H. Schlegel, eds., *The Prokaryotes*, pp. 290–98. Berlin, New York: Springer-Verlag.

Cavalier-Smith, T. 1975. The origin of nuclei and of eukaryotic cells. *Nature* 256: 463–68.

Cavalier-Smith, T. 1987a. The origin of eukaryotic and archaebacterial cells. *Annals of the New York Academy of Sciences* 503: 17–54.

Cavalier-Smith, T. 1987b. The simultaneous symbiotic origin of mitochondria, chloroplasts, and microbodies. *Annals of the New York Academy of Sciences* 503: 55–71.

Clements, K. D. and S. Bullivans. 1991. An unusual symbiont from the gut of surgeon fishes may be the largest known prokaryote. *J. Bact.* 173: 5359–62.

Cleveland, L. R. 1947. The origin and evolution of meiosis. *Science* 105: 287–89.

Cleveland, L. R. 1956. The Flagellates of Termites, Part 3 (16 mm film). Amherst: University of Massachusetts, Zoology.

Cleveland, L. R. and A. V. Grimstone. 1964. The fine structure of the flagellate *Mixotricha* and its associated microorganisms. *Proc. Roy. Soc. London, Ser. B. Biol. Sci.* 159: 668–86.

Coughter, J. and G. Stewart. 1989. Genetic exchange in the environment. *Antonie van Leeuwenhoek* 55: 15–22.

Darnell, J. E. and W. F. Doolittle. 1986. Speculations on the early course of evolution. *PNAS* 83: 1271–75.

Darnell, J., H. Lodish, and D. Baltimore. 1990. *Molecular Cell Biology.* New York: Scientific American Books.

deBoer, P., R. Crossley, and L. Rothfield. 1992. The essential bacterial cell division protein FtsZ is a GTPase. *Nature* 359: 2254–56.

Delias, N. and G. Fox. 1987. Origins of plant chloroplasts and mitochondria based on comparisons of 5S ribosomal RNA. *Annals of the New York Academy of Sciences* 503: 92–102.

dePamphilis, C. W. and J. D. Palmer. 1990. Loss of photosynthetic and respiratory genes from the plastid genomes of a parasitic flowering plant. *Nature* 348: 337–39.

Dippell, R. 1976. Effects of nuclease and protease digestion on the ultrastructure of *Paramecium* basal bodies. *J. Cell Biol.* 69: 622–37.

Dobell, C. 1958. *Anthony van Leeuwenhoek and his 'Little Animacules.'* New York: Russell and Russell.

Doolittle, R. F., D. F. Feng, K. L. Anderson, and M. R. Alberro. 1990. A naturally occurring horizontal gene transfer from eukaryote to prokaryote. *J. Mol. Evol.* 31: 383–88.

Doolittle, W. F. 1987. The evolutionary significance of the archaebacteria. *Annals of the New York Academy of Sciences* 503: 72–77.

Douglas, S. E., C. A. Murphy, D. F. Spencer, and M. W. Gray. 1991. Cryptomonad algae are evolutionary chimeras of two phylogenetically distinct unicellular eukaryotes. *Nature* 350: 148–51.

Dunlop, J., M. Muir, V. Milne, and D. Groves. 1978. A new microfossil assemblage from the Archean of Western Australia. *Nature* 274: 676–78.

Dyer, B. D. 1989a. Symbiosis and organismal boundaries. *American Zoologist* 29: 1085–93.

Dyer, B. D. 1989b. Pyrsonymphida. In L. Margulis, J. Corliss, M. Melkonian, and D. Chapman, eds., *The Handbook of Protoctista,* pp. 266–69. Boston: Jones and Bartlett.

Dyer, B. D. 1989c. *Metopus, Cyclidium* and *Sonderia:* Ciliates enriched and cultured from sulfureta of a microbial mat community. *Biosystems* 23: 41–51.

Dyer, B. D. and R. Obar, eds. 1986. *The Origin of Eukaryotic Cells.* New York: Van Nostrand Reinhold.

REFERENCES

Dyer, T. 1984. The chloroplast genome: Its nature and role in development. In N. R. Baker and J. Barber, eds., *Chloroplast Biogenesis,* pp. 225–69. New York: Elsevier.

Fauré-Fremiet, E. 1951. The marine and sand dwelling ciliates of Cape Cod. *Biol Bull.* 100: 59–70.

Fenchel, T. and B. J. Findlay. 1990a. Endosymbiotic methanogenic bacteria in anaerobic ciliates: Significance for growth efficiency of the host. *J. Protozool.* 38: 18–22.

Fenchel, T. and B. J. Findlay. 1990b. Synchronous division of an endosymbiotic methanogenic bacterium in an anaerobic ciliate *Plagiopyla frontata. J. Protozool.* 38: 22–28.

Fenchel, T., T. Perry, and A. Thane. 1977. Anaerobiosis and symbiosis with bacteria in free-living ciliates. *J. Protozool.* 24: 154–63.

Findlay, B. J. and T. Fenchel. 1989. Hydrogenosomes in some anaerobic protozoa resemble mitochondria. *FEMS Microbiol. Letters* 65: 311–14.

Frankel, J. 1974. Positional information in unicellular organisms. *J. Theor. Biol.* 47: 439–81.

Fuerst, J. A. and R. I. Webb. 1991. Membrane-bounded nucleoid in the eubacterium *Gemmata obscuriglobus. PNAS* 88: 8184–88.

Fulton, C. and A. Dingle. 1971. Basal bodies but not centrioles in *Naeglaria. J. Cell Bio.* 51: 826–36.

Gellissen, G. and G. Michaelis. 1987. Gene transfer from mitochondria to nucleus. *Annals of the New York Academy of Sciences* 503: 391–401.

Gibbs, S. P. 1981. The chloroplasts of some algal groups may have evolved from endosymbiotic eukaryotic algae. *Annals of the New York Academy of Sciences* 361: 193–208.

Gillham, N. W. 1978. *Organelle Heredity.* New York: Raven Press.

Gogarten, J. P., H. Kibak, P. Dittrich, L. Taiz, E. J. Bowman, B. Bowman, M. Manolson, R. Poole, T. Date, T. Oshima, J. Konishi, K. Denda, and M. Yoshida. 1989. Evolution of the vacuolar H+ ATPase: Implications for the origin of eukaryotes. *PNAS* 86: 6661–65.

Golenberg, E. M., D. E. Ginnassi, M. Clegg, C. Smiles, M. Durbin, D. Henderson, and G. Zurawski. 1990. Chloroplast DNA sequences from a Miocene magnolia species. *Nature* 344: 656–58.

Golubic, S. 1992. Stromatolites of Shark Bay. In L. Margulis and L. Olendzenski, eds., *Environmental Evolution.* Cambridge, Mass.: MIT Press.

Gould, S. J. 1991. *Bully for Brontosaurus.* New York: Norton.

Grimes, G. 1976. Laser microbeam induction of incomplete doublets of *Oxytricha fallax. Genet. Res. Camb.* 27: 213–26.

Grimes, G. 1982. Nongenic inheritance: A determinant of cellular architecture. *BioScience* 32: 279–80.

Grimes, G., M. McKenna, C. Goldsmith-Spoegler, and E. Knaupp. 1980. Patterning and assembly of ciliature are independent processes in hypotrich ciliates. *Science* 209: 281–83.

Grosovsky, B. and L. Margulis. 1982. Termite microbial communities. In

R. G. Burns and J. H. Slater, eds., *Experimental Microbial Ecology.* Oxford: Blackwell.

Guerrero, R., I. Esteve, C. Pedrós-Alió, and N. Gaju. 1987. Predatory bacteria in prokaryotic communities. *Annals of the New York Academy of Sciences* 503: 238–50.

Gupta, K. P., K. L. van Golen, E. Randerath, and K. Randerath. 1990. Age-dependent covalent DNA alterations (I-compounds) in rat liver mitochondrial DNA. *Mutat. Res.* 237: 17–27.

Gyllensten, U., D. Wharton, A. Josefsson, and A. Wilson. 1991. Paternal inheritance of mitochondrial DNA in mice. *Nature* 352: 255–57.

Hagemann, S., W. Miller, and W. Pinsker. 1990. P-related sequences in *Drosophila bifasciata:* A molecular clue to the understanding of P-element evolution in the genus *Drosophila. J. Mol. Evol.* 31: 478–84.

Hall, J., Z. Ramanis, and D. Luck. 1989. Basal body/centriolar DNA: Molecular genetics studies in *Chlamydomonas. Cell* 59: 121–32.

Hamilton, W. D., R. Axelrod, and R. Tenese. 1990. Sexual reproduction as an adaptation to resist parasites. *PNAS* 87: 3566–73.

Hammersmith, R. and G. Grimes. 1981. Effects of cystment on cells of *Oxytricha fallax* possessing supernumerary dorsal bristle rows. *J. Embryol. Exp. Morph.* 63: 17–27.

Han, T. M. and B. Runnegar. 1992. Megascopic eukaryotic algae from the 2.1 billion year old Negaunee Iron Formation Michigan. *Science* 257: 232–35.

Hartman, J. M., J. P. Puma, and T. Gurney. 1974. Evidence for the association of RNA with ciliary basal bodies of *Tetrahymena. J. Cell Sci.* 16: 241–60.

Hartwell, L. H. 1976. Sequential function of gene products relative to DNA synthesis in the yeast cell cycle. *J. Mol. Biol.* 104: 803–17.

Hayes, J. M. 1983. Geochemical evidence bearing on the origin of anaerobiosis, a speculative hypothesis. In J. W. Schopf, ed., *Earth's Earliest Biosphere.* Princeton, N.J.: Princeton University Press.

Hayes, J. M., D. Desmarais, I. Lambert, and H. Sterns. 1992. Unsolved problems and conclusions for proterozoic biogeochemistry. In J. W. Schopf amd C. Klein, eds., *The Proterozoic Biosphere,* pp. 133–34. Cambridge: Cambridge University Press.

Heidemann, S. R. and M. W. Kirschner. 1975. Aster formation in eggs of *Xenopus laevis:* Induction by isolated basal bodies. *J. Cell Biol.* 67: 105–17.

Heidemann, S. R., C. R. Sander, and M. Kirschner. 1977. Evidence for a functional role of RNA in centrioles. *Cell* 10: 337–50.

Heinemann, J. and G. Sprague. 1989. Bacterial conjugative plasmids mobilize DNA transfer between bacteria and yeast. *Nature* 340: 205–9.

Hixon, W. G. and D. C. Searcy. 1991. Cytoskeletal apparatus in a thermophilic archaebacterium. Abstract for Cell Biology Meeting, Boston.

Hoffman, E. 1965. The nucleic acids of basal bodies isolated from *Tetrahymena pyriformis. J. Cell Bio.* 25: 217–28.

Hoffman, H. P. and C. Avers. 1973. Mitochondrion of yeast: Ultrastructure evidence for one giant, branched organelle per cell. *Science* 181: 749–51.

Hofmann, H. J. and J. Chen 1981. Carbonaceous megafossils from the Precambrian (1800 Ma) near Jixian, northern China. *Can. J. Earth Sci.* 18: 443–47.

Hofmann, H. J. and J. W. Schopf. 1983. Early proterozoic microfossils. In J. W. Schopf, ed., *Earth's Earliest Biosphere.* Princeton, N.J.: Princeton University Press.

Hofstadter, D. 1985. *Metamagical Themas: Questing for the Essence of Mind and Pattern.* New York: Basic Books.

Horodyski, R. J. 1980. Middle proterozoic shale-facies microbiota from the lower belt supergroup, Little Belt Mountains, Montana. *J. Paleont.* 54: 649–63.

Houck, M., J. Clark, K. Peterson, and M. Kidwell. 1991. Possible horizontal transfer of Drosophila genes by the mite *Proctolaelaps regalis. Science* 253: 1125–29.

Hovind-Hougen, K. 1976. Determination by means of electron microscopy of morphological criteria of value for classification of some spirochetes in particular treponemes. *Acta Pathol. Microbiol. Scand. Sect. B.* Suppl. 255.

Howe, C. J. and A. G. Smith. 1991. Plants without chlorophyll. *Nature* 349: 109.

Huang, B., Z. Ramanis, S. Dutcher, and D. Luck. 1982. Uniflagellar mutant of *Chlamydomonas:* Evidence for the role of basal bodies in transmission of positional information. *Cell* 29: 745–53.

Hufnagel, L. and R. Torch. 1967. Intraclonal dimorphism of caudal cirri in Euplotes vannus: Cortical determination. *J. Protozool.* 14: 429–39.

James, S. W., L. Ranum, C. Silflow, and P. Lefebvre. 1988. Mutants resistant to antimicrotubule herbicide map to a locus on the uni-linkage group in *Chlamydomonas reinhardtii. Genetics* 118: 141–47.

Jeon, K. 1991. Amoeba and X bacteria: Symbiont acquisition and possible species change. In L. Margulis and R. Fester, eds., *Symbiosis as a Source of Evolutionary Innovation.* Cambridge, Mass.: MIT Press.

John, P. 1987. *Paracoccus* as a free-living mitochondrion. *Annals of the New York Academy of Sciences* 503: 140–51.

John, P. and F. R. Whatley. 1975. *Paracoccus denitrificans* and the evolutionary origin of the mitochondrion. *Nature* 254: 495–98.

Johnson, D. and S. Dutcher. 1991. Molecular studies of linkage group XIX of *Chlamydomonas reinhardtii:* Evidence against a basal body location. *J. Cell Biol.* 113: 339–46.

Johnson, K. and J. Rosenbaum. 1990. The basal bodies of *Chlamydomonas reinhardtii* do not contain immunologically detectable DNA. *Cell* 62: 615–19.

Juszczak, A., S. Aono, and M. W. W. Adams. 1991. The extremely thermophilic eubacterium *Thermatoga maritima* contains a novel iron-hydrogenase whose cellular activity is dependent upon tungsten. *J. Biol. Chem.* 266: 13834–41.

Kandler, O. 1993 (in press). The early diversification of life. In S. Bengtsen, ed., *Early Life on Earth.* New York: Columbia University Press.

Kasting, J., H. Holland, and L. Kump. 1992. Atmospheric evolution: The rise of oxygen. In J. W. Schopf and C. Klein, eds., *The Proterozoic Biosphere.* Cambridge: Cambridge University Press.

Kazmierczak, J. 1979. The eukaryotic nature of *Eosphaera*-like ferriferous structures from the precambrian Gunflint iron formation, Canada: A comparative study. *Precambrian Research* 9: 1–22.

Keller, E. F. 1983. *A Feeling for the Organism.* New York: W. H. Freeman.

Kempe, S., J. Kazmierczak, A. Lipp, G. Landman, T. Konuk, and T. Reimer. 1991. Largest known microbiolites discovered in Lake Van, Turkey. *Nature* 349: 605–8.

Kendrick, B. 1985. *The Fifth Kingdom.* Ontario: Mycologue Publications.

Klein, C. 1992. Introduction to the proterozoic atmosphere and ocean. In J. W. Schopf and C. Klein, eds., *The Proterozoic Biosphere.* Cambridge: Cambridge University Press.

Klenk, H. P., P. Palm, F. Lottspeich, and W. Zellig. 1992. Component H of the DNA-dependent RNA polymerase of archaea is homologous to a subunit shared by the three eukaryotic nuclear RNA polymerases. *PNAS* 89: 407–10.

Knoll, A. 1985. Exceptional preservation of photosynthetic organisms in silicified carbonates of silicified peats. *Phil. Trans. Roy. Soc. Lond., Ser. B.* 311: 111–22.

Knoll, A. 1992. The early evolution of eukaryotes: A geological perspective. *Science* 256: 622–27.

Knoll, A. and S. Awramik. 1983. Ancient microbial ecosystems. In W. Krumbein, ed., *Microbial Geochemistry.* Oxford: Blackwell Scientific.

Kraut, M., I. Hugendieck, S. Herwig, and O Meyer. 1989. Homology and distribution of CO dehydrogenase structural genes in carboxydotrophic bacteria. *Archives of Microbiol.* 152: 335–41.

Kusky, T. M. and J. P. Vanyo. 1991. Plate reconstructions using stromatolitic heliotrophism. *J. of Geol.* 99: 321–36.

Kvam, E. T., T. Stokke, and J. Moan. 1990. The lengths of DNA fragments light-induced in the presence of a photosynthesizer localized at the nuclear membrane of human cells. *Biochim. Biophys. Acta* 1049: 33–37.

Lake, J. A. 1988. Origin of the eukaryotic nucleus determined by the rate-invariant analysis of rRNA sequences. *Nature* 331: 184–86.

Lake, J. A. 1989. Origin of the eukaryotic nucleus: Eukaryotes and eocytes are genotypically related. *Can. J. Microbiol.* 35: 109–18.

Landan, G., G. Cohen, Y. Aharonowitz, Y. Shuali, D. Graur, and D. Shiffman. 1990. Evolution of isopenicillin N synthase genes may have involved horizontal gene transfer. *Mol. Biol. Evol.* 7: 399–406.

Lee, J. J. 1989. Granuloreticulosa. In L. Margulis, J. Corliss, M. Melkonian, and D. Chapman, eds., *Handbook of Protoctists,* pp. 524–48. Boston: Jones and Bartlett.

Lee, Y., D. Friedman, and F. Ayala. 1985. Superoxide dismutase: An evolutionary puzzle. *PNAS* 82: 824–28.

Lefort-Tran, M. 1983. Phylogeny and organization of chloroplast envelopes. *Endocytobiology II.* New York: W. deGruyter.

Lewin, B. 1987. *Genes III.* New York: Wiley.

Lewin, R. 1981. The Prochlorophytes. In M. Starr, H. Stolp, H. Trüper, A. Balows, and H. Schlegel, eds., *The Prokaryotes,* pp. 257–66. Berlin: Springer-Verlag.

Licari, G. R. and P. E. Cloud. 1968. Reproductive structures and taxonomic affinities of some nanofossils from the Gunflint iron formation. *PNAS* 59: 1053–60.

Margulis, L. 1981. *Symbiosis in Cell Evolution.* San Francisco: W. H. Freeman.

Margulis, L. 1982. *Early Life.* Boston: Science Books International.

Margulis, L. 1991. Symbiogenesis and symbionticism. In L. Margulis and R. Fester, eds., *Symbiosis as a Source of Evolutionary Innovation.* Cambridge, Mass.: MIT Press.

Margulis, L. 1993. *Symbiosis in Cell Evolution.* Boston: W. H. Freeman.

Margulis, L. and D. Sagan. 1986. *Origins of Sex.* New Haven: Yale University Press.

Margulis, L. and K. Schwartz. 1988. *Five Kingdoms.* San Francisco: W. H. Freeman.

Margulis, L., L. To, and D. Chase 1978. Microtubules in prokaryotes. *Science* 200: 1118–24.

Margulis, L., J. Corliss, M. Melkonian, and D. Chapman, eds. 1989. *Handbook of Protoctista.* Boston: Jones and Bartlett.

Masover, G. and L. Hayflick. 1981. The genera *Mycoplasma, Ureaplasma* and *Acholeplasma* and associated organisms. In M. Starr, H. Stolp, H. Trüper, A. Balows, and H. Schlegel, eds., *The Prokaryotes.* Berlin: Springer-Verlag.

McLaughlin, J. and P. Zahl. 1966. Endozoic algae. In S. M. Henry, ed., *Symbiosis,* vol. 1, pp. 257–97. New York: Academic Press.

Mendelson, C. and J. W. Schopf. 1992. Proterozoic and early Cambrian acritarchs. In J. W. Schopf and C. Klein, eds., *The Proterozoic Biosphere.* Cambridge: Cambridge University Press.

Mereschkowsky, C. 1910. Theorie der zwei Plasmaarten als Grundlage der Symbiogenesis einer neuen Lehre von der Entstehung der Organismen. Cited by F. J. R. Taylor. 1987. An overview of the status of evolutionary cell symbiosis theories. *Annals of the New York Academy of Sciences* 503: 1–16.

Mindlin, S. Z., Zh. Gorlenko, I. Bass, and N. Khachikian. 1990. Spontaneous transformation in mixed cultures of various types of Acinetobacter and during joint growth of *Acinetobacter calcoaceticus* with *Escherischia coli* and *Pseudomonas aereuginosa. Genetika* 26: 1729–39 (in Russian).

Moll, R. and G. Schaefer. 1991. Purification and characterization of an archaebacterial succinate dehydrogenase complex from the plasma membrane of the thermoacidophile *Sulfulobus acidocaldarius. Eur. J. Biochem.* 201: 593–600.

Muller, M. 1988. Energy metabolism of protozoa without mitochondria. *Ann. Rev. Microbiol.* 42: 465–88.

Murray, A. and J. Szostak. 1985. Chromosome segregation in mitosis and meiosis. *Ann. Rev. Cell Biol.* 1: 289–315.

Muscatine L. and R. R. Pool. 1979. Regulation of numbers of intracellular algae. *Proc. Roy. Soc. London, Ser. B* 204: 131–39.

Muskhelishvili, G. D., M. V. Karseladze, and D. A. Prangishvili. 1990. Sensitivity to steroids and steroid binding zones of thermoacidophilic archaebacterium *Sulfulobus acidocalderius. Doklady Akade, Nauk* 313: 1259–62 (abstract).

Ninan, C. A. 1958. (as cited in Stebbins 1971).

Nixon, A. and P. R. Norris. 1992. Autotrophic growth and inorganic sulphur compound oxidation of *Sulfulobus* in chemostat culture. *Archives of Microbiol.* 157: 155–60.

Nugent, J. M. and J. D. Palmer. 1991 RNA-mediated transfer of the gene coxII from the mitochondrion to the nucleus during flowering plant evolution. *Cell* 66: 473–81.

Obar, R. 1993. GTP-binding in bacterial septation. *Trends in Cell Biology* 3: 4.

Obar, R. and J. Green. 1985. Molecular archaeology of the mitochondrial genome. *J. Mol. Evol.* 22: 243–51.

Oehler, D. Z. 1976. Transmission electron microscopy of organic microfossils from the late precambrian Bitter Springs formation of Australia. *J. Paleont.* 50: 90–106.

Opperdoes, F. R. and P. A. M. Michels. 1989. Biogenesis and evolutionary origin of peroxisomes. In J. M. Tager, A. Azzi, S. Papa, and F. Guerri, eds., *Organelles in Eukaryotic Cells.* New York: Plenum Press.

Palazzo, R. E., E. Vaisberg, R. W. Cole, and C. L. Rieder. 1992. Centriole duplication in lysates of *Spisula solidissima* oocytes. *Science* 256: 219–21.

Palenick, B. and R. Haselkorn. 1992. Multiple evolutionary origins of prochlorophytes, the chlorophyll b containing prokaryotes. *Nature* 355: 265–67.

Palmer, J. D. 1992. Green ancestry of malarial parasites? *Curr. Biol.* 2: 318–20.

Palmer, J. D. and J. M. Logsdon, Jr. 1991. The recent origins of introns. *Current Opinions in Genetics and Devel.* 1: 470–77.

REFERENCES

Peñalva, M., A. Moya, J. Dopazo, and D. Ramon. 1990. Sequences of isopenicillin N synthetase genes suggest horizontal gene transfer from prokaryotes to eukaryotes. *Proc. Roy. Soc. Lond., Ser. B* 241: 164–69.

Perler, F., D. Comb, W. Jack, L. Moran, B. Quiang, R. Kucera, J. Benner, B. Slatko, D. Nwankwo, S. Hempstead, C. Carlow, and H. Jannosch. 1992. Intervening sequences in an Archaea DNA polymerase gene. *PNAS* 89: 5577–81.

Pickett-Heaps, J. D. 1971. The autonomy of the centriole: Fact or fallacy. *Cytobios* 3: 205–14.

Plos, K. S., I. Hull, B. R. Levin, I. Orskov, F. Orskov, and C. Svanborg-Eden. 1989. Distribution of the p-associated pilus (pap) region among *Escherischia coli* from natural sources: Evidence for horizontal gene transfer. *Infect. and Immunit.* 57: 1604–11.

Porter, K. R. and J. B. Tucker. 1981. The ground substance of the living cell. *Sci. Amer.* 244: 56–67.

Prebble, J. N. 1981. *Mitochondria, Chloroplasts, and Bacterial Membranes.* London: Longman.

Raikov, I. B. 1982. *The Protozoan Nucleus.* New York: Springer-Verlag.

Ramanis, Z. and D. Luck. 1986. Loci affecting flagellar assembly and function map to an unusual linkage group in *Chlamydomonas reinhardtii. PNAS* 83: 436–38.

Randall, J. T. and C. Disbrey. 1965. Evidence for the presence of DNA at basal body sites in *Tetrahymena pyriformis. Proc. Roy. Soc. Lond., Ser. B* 162: 473–491.

Randall, J. T. and J. M. Hopkins. 1963. Studies of cilia, basal bodies and related organelles II: Problems of genesis. *Proc. Linn. Soc. of Lond., Ser. B* 174: 37–39.

RayChaudhuri, D. and J. T. Park. 1992. *Escherischia coli* cell division gene ftsZ encodes a novel GTP-binding protein. *Nature* 359: 251–54.

Reiter, W. D., U. Huedepohl, W. Zillig. 1990. Mutational analysis of an archaebacterial promotor: Essential role of a TATA box for transcription efficiency and start-site selection in vitro. *PNAS* 87: 9509–13.

Ripley, M. and A. Anilionis. 1978. Evolution of the bacterial genome. *Ann Rev. Mic.* 32: 519–60.

Rivera, M. and J. Lake. 1992. Evidence that eukaryotes and eocyte prokaryotes are immediate relatives. *Science* 257: 74–76.

Sagan, L. (Margulis). 1967. On the origin of mitosing cells. *J. Theor. Biol.* 14: 225–74.

Sandman, K., J. Krzyski, B. Dobrinski, R. Lurz, and J. Reeve. 1990. Hmf, a DNA-bindinng protein isolated from the hyperthermophilic archaeon *Methanothermus fervidus* is most closely related to histones. *PNAS* 87: 5788–91.

Sangwan, I. and M. O'Brian. 1991. Evidence for an interorganismic heme biosynthetic pathway in symbiotic soybean root nodules. *Science* 251: 1220–22.

Schenkinger, M. F., B. Redl, and G. Stoeffler. 1991. Purification and properties of an extreme glutamate dehydrogenase from the archaebacterium *Sulfulobus solfataricus. Biochem. and Biophys. Acta* 1073: 142–48.

Schidlowski, M., J. M. Hayes, and I. R. Kaplan. 1983. Isotopic inferences of ancient biochemistries: Carbon, sulfur, hydrogen and nitrogen. In J. W. Schopf, ed., *Earth's Earliest Biosphere.* Princeton, N.J.: Princeton University Press.

Schlegel, H. and H. W. Jannasch. 1981. Prokaryotes and their habitats. In M. Starr, H. Trüper, A. Balows, and H. Schlegel, eds., *The Prokaryotes.* Berlin: Springer-Verlag.

Schopf, J. W. 1968. Microflora of the Bitter Springs Formation Late Precambrian, Central Australia. *J. of Paleontology* 42: 651–88.

Schopf, J. W. 1992. Paleobiology of the archaean. In J. W. Schopf and C. Klein, eds., *The Proterozoic Biosphere,* pp. 25–39. Cambridge: Cambridge University Press.

Schopf, J. W. and D. Z. Oehler. 1976. How old are the eukaryotes? *Science* 193: 47–49.

Schopf, J. W. and M. R. Walter. 1983. Archean microfossils: New evidence of ancient microbes. In J. W. Schopf, ed., *Earth's Earliest Biosphere.* Princeton, N.J.: Princeton University Press.

Schuster, W., R. Hiesel, B. Wissinger, and A. Brennicke. 1990. RNA editing in the cytochrome b locus of the higher plant *Oenothera berteriana* includes a U to C transition. *Mol. and Cell Bio.* 10: 2428–31.

Seaman, G. R. 1959. Large scale isolation of kinetosomes from the ciliated protozoan *Tetrahymena pyriformis. Exp. Cell Res.* 21: 292–302.

Searcy, D. G., D. B. Stein, and K. B. Searcy. 1981. A Mycoplasma-like archaebacterium possibly related to the nucleus and cytoplasm of eukaryotic cells. *Annals of the New York Academy of Sciences* 361: 312–24.

Searcy, D. G. and F. R. Whatley. 1982. *Thermoplasma acidophilum* cell membrane: Cytochrome b and sulfate stimulated ATPase. *Zbl. Bact. Hyg. I. Abt. Orig.* C3: 245–57.

Searcy, D. G. and F. R. Whatley. 1984. *Thermoplasma acidophilum:* Glucose degradative pathways and respiratory activities. *Syst. Appl. Microbiol.* 5: 30–40.

Shivley, J. M. 1974. Inclusion bodies of prokaryotes. *Ann. Rev. Micro.* 28: 167–87.

Siddall, M. E., H. Hong, and S. Desser. 1992. Phylogenetic analysis of Diplomonida: Evidence for heterochrony in protozoa and against *Giardia lamblia* as a "missing link." *J. Protozool.* 39: 361–67.

Siemeister, G. and W. Hachtel. 1989. A circular 73 kb DNA from the colorless flagellate *Astasia longa* that resembles the chloroplast DNA of *Euglena. Curr. Genet.* 15: 435–42.

Sioud, M., G. Baldacci, P. Forterre, and A. M. De Recondo. 1987. Antitumor drugs inhibit the growth of halophilic archaebacteria. *Eur. J. of Biochem.* 169: 231–36.

Sluder, G. 1989. Centrosomes and the cell cycle. *J. Cell Sci. Suppl.* 12: 253–75.

Sluder, G., F. J. Miller, and C. L. Rieder. 1989. Reproductive capacity of sea urchin centrosomes without centrioles. *Cell Motility and the Cytoskeleton* 13: 264–73.

Sluder, G., F. J. Miller, K. Lewis, E. D. Davis, and C. L. Rieder. 1989. Centrosome inheritance in starfish zygotes: Selective loss of the maternal centrosome after fertilization. *Dev. Bio.* 131: 567–79.

Smith, D. C. 1979. The establishment of a symbiosis. *Proc. Roy. Soc. Lond., Ser. B.* 204: 115–30.

Sogin, M. 1991. Early evolution and the origin of eukaryotes. *Curr. Opin. in Gen. and Dev.* 1: 457–63.

Sogin, M. L., J. H. Gunderson, H. E. Elwood, R. A. Alonso, and D. Peattie. 1989. Phylogenetic meaning of the kingdom concept: An unusual rRNA from *Giardia lamblia. Science* 243: 75–77.

Sonea, S. and M. Panisset. 1983. *A New Bacteriology.* Boston: Jones and Bartlett.

Sonneborn, T. M. 1963. Does preformed cell structure play an essential role in cell heredity? In J. M. Allen, ed., *The Nature of Biological Diversity,* pp. 161–221. New York: McGraw Hill.

Sonneborn, T. M. 1970. Determination, development and inheritance of the structure of the cell cortex. In H. Padykula, ed., *Control Mechanisms in the Expression of Cellular Phenotypes,* pp. 1–13. New York: Academic Press.

Stachel, S. and P. Zambryski. 1989. Generic trans-kingdom sex? *Nature* 340: 190–91.

Stack, S. and W. Brown. 1969. Somatic pairing reduction and recombination: An evolutionary hypothesis of meiosis. *Nature* 222: 1275–76.

Stanier, R. Y., N. Pfennig, and H. Trüper. 1981. Introduction to phototrophic prokaryotes. In M. P. Starr, H. Stolp, H. G. Trüper, A. Balows, and H. G. Schlegel, eds., *The Prokaryotes,* p. 197–211. Berlin: Springer-Verlag.

Starr, M. P. and J. M. Schmidt. 1981. Prokaryotic diversity. In M. P. Starr, H. Stolp, H. G. Trüper, A. Balows, and H. G. Schlegel, eds., *The Prokaryotes.* Berlin: Springer-Verlag.

Starr, M. P., H. Stolp, H. G. Trüper, A. Balows, and H. G. Schlegel, eds. 1981. *The Prokaryotes.* Berlin: Springer-Verlag.

Stebbins, G. L. 1971. *Chromosomal Evolution in Higher Plants.* Reading, Mass.: Addison-Wesley.

Stern, D. B. and D. M. Lonsdale. 1982. Mitochondrial and chloroplast genomes of maize have a 12-kilobase DNA sequence in common. *Nature* 299: 698–702.

Summons, R. E. and M. R. Walter. 1990. Molecular fossils and microfossils of prokaryotes and protists from proterozoic sediments. *Am. J. of Science* 290A: 212.

Syvanen, M. 1987. Molecular clocks and evolutionary relationships: Possible distortions due to horizontal gene flow. *J. Mol. Evol.* 26: 16–23.

Tabak, H. F. and B. Distal. 1989. Biogenesis of peroxisomes. In J. M. Tager, A. Azzi, S. Papa, and F. Guerreri, eds., *Organelles in Eukaryotic Cells.* New York: Plenum Press.

Tamm, S. 1982. Flagellated ectosymbiotic bacteria propel a eukaryotic cell. *J. Cell Biol.* 94: 697–709.

Tappan, H. 1976. Possible eukaryotic algae (Bangiophycidae) among early proterozoic microfossils. *GSA Bulletin* 87: 633–39.

Taylor, F. J. R. 1983. Some eco-evolutionary aspects of intracellular symbioses. *Int. Rev. Cyt. Suppl.* 14: 1–25.

Thorsness, P. and T. Fox. 1990. Escape of DNA from mitochondria to the nucleus in *Saccharomyces cerevisiae. Nature* 346: 376–79.

Urbach, E., D. Robertson, and S. Chisholm. 1992. Multiple evolutionary origins of prochlorophytes within cyanobacterial radiation. *Nature* 355: 267–69.

van den Boogaart, P., J. Samallo, and E. Agsterribe. 1982. Similar genes for mitochondrial ATPase subunit in the nuclear and mitochondrial genomes of *Neurospora crassa. Nature* 298: 187–89.

Vickermann, K. 1989. Diplomonida. In L. Margulis, J. Corliss, M. Melkonian, and D. Chapman, eds., *The Handbook of Protoctists,* pp. 200–10. Boston: Jones and Bartlett.

Vidal, G. 1984. The oldest eukaryotic cells. *Sci. Amer.* 250: 48–57.

Vitaya, V. B. and K. Toda. 1991. Physiological adsorption of *Sulfolobus acidiocalderius* on coal surfaces. *Appl. Micro. and Tech.* 35: 690–95.

Volker, A., J. Wolters, T. Pieler, M. Digweed, T. Spech, and N. Ulbrich. 1987. Evolution of organisms and organelles as studied by comparative computer and biochemical analysis of ribosomal 5S RNA. *Annals of the New York Academy of Sciences* 503: 103–24.

Wächtershäuser, G. 1990. Evolution of the first metabolic cycles. *PNAS* 87: 200–204.

Walker, J. 1977. *The Flying Circus of Physics with Answers.* New York: Wiley.

Walker, J., C. Klein, M. Schidlowski, J. W. Schopf, D. Stevenson, and M. Walker. 1983. Environmental evolution of the Archaean–Early Proterozoic Earth. In J. W. Schopf, ed., *Earth's Earliest Biosphere.* Princeton, N.J.: Princeton University Press.

Wallin, I. E. 1923. The mitochondria problem. *Am. Nat.* 57: 255–61.

Wallsgrove, R. M. 1991. Plastid genes and parasitic plants. *Nature* 350: 664.

Walter, M. R. 1983. Archean stromatolites: Evidence of the earth's earliest benthos. In J. W. Schopf, ed., *Earth's Earliest Biosphere.* Princeton, N.J.: Princeton University Press.

Walter, M. R., D. Rulin, R. J. Horodyski. 1990. Coiled carbonaceous megafossils from the middle Proterozoic Jixian (Tianjin) and Montana. *Am. J. Sci.* 290-A: 133–48.

REFERENCES

Weigel, B. J., S. G. Burgett, V. Chen, P. Skatrud, C. Frolik, S. Queener, T. Ingolia. 1988. Cloning and expression in *Escherischia coli* of isopenicillin N-synthetase genes from *Streptomyces lipmanii* and *Aspergillus Nidulans. J. Bact.* 170: 3817–26.

Weihs, D. 1973. Hydromechanics of fish schooling. *Nature* 241: 290.

Whatley, J. M. and C. Chapman-Anderson. 1989. Karyoblastea. In L. Margulis, J. Corliss, M. Melkonian, and D. Chapman, eds., *The Handbook of Protoctists,* pp. 167–85. Boston: Jones and Bartlett.

Whatley, F. R. and J. M. Whatley. 1983. *Pelomyxa palustris. Endocytobiology II.* Berlin: Walter de Gruyter.

Wheatley, D. N. 1982. *The Centriole: A Central Enigma in Cell Biology.* Amsterdam: Elsevier.

Woese, C. R. 1987. Bacterial evolution. *Microbiol. Rev.* 51: 221–71.

Woese, C. R. and G. E. Fox. 1977. Phylogenetic structure of the prokaryotic domain: The primary kingdoms. *PNAS* 74: 5088–90.

Woese, C. R., J. Gibson, and G. E. Fox. 1980. Do genealogical patterns in purple photosynthetic bacteria reflect interspecific gene transfer? *Nature* 283: 212–14.

Woese, C. R., O. Kandler, and M. Wheelis. 1990. Towards a natural system of organisms: Proposal for the domains Archea, Bacteria and Eucarya. *PNAS* 87: 4576–79.

Woese, C. R., E. Stackebrandt, and W. Ludwig. 1985. What are mycoplasmas: The relationship of tempo and mode in bacterial evolution. *J. Mol. Evol.* 21: 305–16.

Wolfe, K. H., C. W. Morden, and J. D. Palmer. 1992. Small single copy region of plastid DNA in the non-photosynthetic angiosperm *Epifagus virginiana* contains only two genes. *J. Mol. Biol.* 223: 95–104.

Wolters, J. 1991. The troublesome parasites—molecular and evolutionary evidence that apicomplexa belong to the dinoflagellate-ciliate clade. *Biosystems* 25: 75–83.

Yamao, F., A. Muto, Y. Kawauchi, M. Iwami, S. Iwagami, Y. Azumi, and S. Osawa. 1985. UGA is read as tryptophan in *Mycoplasma capricolum. PNAS* 82: 2306–9.

Younger, K. B., S. Banerjee, J. K. Kelleher, M. Winston, and L. Margulis. 1972. Evidence that the synchronized production of new basal bodies is not associated with DNA synthesis in *Stentor coeruleus. J. Cell Sci.* 11: 1–17.

Zhang, Zh. 1986. Clastic facies microfossils from Chuanlinggou Formation (1800ma) near Jixian, North China. *J. Micropal.* 5: 9–16.

Zhore, D. and R. H. White. 1991. Transsulfuration of archaebacteria. *J. Bact.* 173: 3250-51.

INDEX

INDEX

INDEX